The identification of progress in learning

The European Science Foundation is an international non-governmental organization with its seat in Strasbourg (France). Members are Academies and Research Councils which are responsible for supporting scientific research at a national level, and which are funded largely from government sources. The term 'science' is used in its broadest sense to include the humanities, social sciences, biomedical sciences and the natural sciences with mathematics. The ESF has currently 48 members from 18 European countries.

The tasks of the ESF are to:

assist its Member Organizations to coordinate their research programmes and define their priorities;

identify areas in need of stimulation, particularly those of an interdisciplinary nature;

further cooperation between researchers by facilitating their movement between laboratories, holding workshops, managing support schemes approved by the Member Organizations, and arranging for the joint use of special equipment;

harmonize and assemble data needed by the Member Organizations;

foster the efficient dissemination of information;

respond to initiatives which are aimed at advancing European science;

maintain constructive relations with the European Communities and other relevant organizations.

The ESF is funded through a general budget to which all Member Organizations contribute (according to a scale assessed by country), and a series of special budgets covering Additional Activities not included in the main programme, to which only participating Organizations contribute. The programmes of the ESF are determined by the Assembly of all Member Organizations. Their implementation is supervised by an elected Executive Council, assisted by the Office of the Foundation which consists of an international staff directed by the Secretary General and located in Strasbourg.

The identification of progress in learning

EDITED BY
T. HÄGERSTRAND

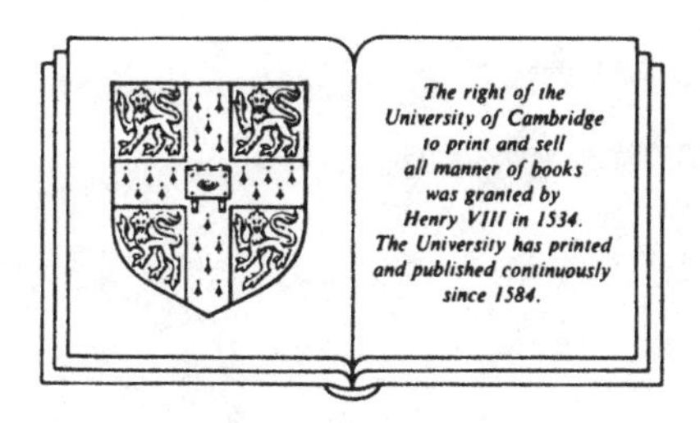
The right of the
University of Cambridge
to print and sell
all manner of books
was granted by
Henry VIII in 1534.
The University has printed
and published continuously
since 1584.

CAMBRIDGE UNIVERSITY PRESS
CAMBRIDGE
LONDON NEW YORK NEW ROCHELLE
MELBOURNE SYDNEY

CAMBRIDGE UNIVERSITY PRESS
Cambridge, New York, Melbourne, Madrid, Cape Town, Singapore, São Paulo, Delhi

Cambridge University Press
The Edinburgh Building, Cambridge CB2 8RU, UK

Published in the United States of America by Cambridge University Press, New York

www.cambridge.org
Information on this title: www.cambridge.org/9780521106085

First published 1985
This digitally printed version 2009

A catalogue record for this publication is available from the British Library

Library of Congress Catalogue Card Number: 84–14277

ISBN 978-0-521-30087-2 hardback
ISBN 978-0-521-10608-5 paperback

Contents

Contributors

Sir Michael Atiyah, FRS
University of Oxford
Mathematical Institute
24–29 St Giles
Oxford OX1 3LB, UK

Professor R. Boudon
GEMAS
Maison des Sciences de l'Homme
54 boulevard Raspail
75270 Paris, France

Professor R. B. Clark
University of Newcastle upon Tyne
Department of Zoology
The University
Newcastle upon Tyne NE1 7RU, UK

Professor S. C. Dik
University of Amsterdam
Department of Linguistics
Spuistraat 210 Postbus 19188
Amsterdam 1000 GD, Netherlands

Professor W. U. Dressler
University of Vienna
Dr Karl-Lueger-Ring 1
1010 Vienna, Austria

Professor H. R. Duncker
Institut für Anatomie
Aulweg 123
D-6300 Giessen, West Germany

Professor S. N. Eisenstadt
The Hebrew University of Jerusalem
Faculty of Social Sciences
Mount Scopus
91 905 Jerusalem, Israel

Professor H. G. Gyllenberg
Department of Microbiology
University of Helsinki
00710 Helsinki 71, Finland

Professor H. Haken
Institut für Theoretische Physik
Universität Stuttgart
Pfaffenwaldring 57/IV
7000 Stuttgart 80, West Germany

Professor Inge Jonsson
Litteraturvetenskapliga Institutionen
Stockholms Universitet
S-106 91, Sweden

Professor G. Kauffmann
Institut für Kunstgeschichte
Domplatz 23
4400 Münster, West Germany

Professor E. Malinvaud
42 avenue de Saxe
75007 Paris, France

Professor Helga Nowotny
Vienna Center
Berggasse 17
1090 Vienna, Austria

Professor L. Pasinetti
Università Catolica
Largo Gemelli 1
Milan 20123, Italy

Professor C. Rizzuto
Università Di Genova
Instituto Di Scienze Fisiche
Viale Bendetto XV, 5
Genova 16132, Italy

Professor L. B. Slobodkin
State University of New York at Stony Brook
Department of Ecology & Evolution
Stony Brook
New York, 11790, USA

Professor R. Thom
Institut des Hautes Etudes Scientifiques
35 route de Châtres
91440 Bures sur Yvette, France

Professor H. L. Wesseling
Rijksuniversiteit Te Leiden
Faculteit der Letteren
Doelensteeg 16 – 2311 VL
Postbus 9515 2300 RA
Leiden, Netherlands

Dr H. R. Wulff
Consultant Physician
Herlev University Hospital
Medical Department C
DK-2730 Herlev, Denmark

Professor J. M. Ziman, FRS
Imperial College of Science and Technology
Department of Social and Economic Studies
53 Prince's Gate
Exhibition Road
London SW7 2PG, UK

Preface

'The identification of progress in learning' was the theme of an international and multidisciplinary colloquium organized by the European Science Foundation in Colmar, France, on 24–26 March 1983. This volume contains the key-note address, the written contributions, the concluding reflections and the separate comments which introduced the discussions of some 55 participants representing a broad range of disciplines and nations.

Debating the nature of scientific knowledge and the meaning of scientific advancement has a long tradition among philosophers and historians of science. These scholars have to a large extent preferred to study past achievements and have tended to concentrate their attention on a very restricted sector of science.

However, in addition to the very general syntheses they have put forward, there exists a rich source of insight in the assessments of the progress of students engaged in actual scientific practice. Such analyses are largely hidden within the multiplicity of specialities which today make up the scientific scene. Busy supervisors have little time to express their views on these matters, but their silence does not mean they have no opinions. In practice, the criteria needed to distinguish between the promising and the less promising steps to take exist and are employed to guide the choice of problems and methods as well as to evaluate the work of colleagues and pupils when it comes to funding, publication and promotion. But it seems as if such criteria are, in general, more implicit than explicit. They are part of the 'taken-for-granted' culture of the separate fields of learning and they are elaborated through a mute agreement between those who have been trained in a certain tradition.

There are external dangers as well as internal disadvantages in the situation. Especially when the economic conditions are strained as now, it is essential to be able to explain clearly to the outside world one's system of priorities and it is then easier to pick on immediate practical goals rather than time-consuming, free-ranging and innovative research. At a time of increasing specialization it is, furthermore, essential to try to overcome the potential mutual isolation of specialists which runs counter to the interests of reciprocal understanding and cross-fertilization of ideas. This, in turn, hampers the development of learning as a collective human endeavour.

The European Science Foundation is involved in the cooperation of all fields of learning from the humanities and social sciences to the medical, natural and technical sciences. For an organization with this exceptionally broad scope it is an obvious task to try to stimulate scientists to reflect upon their priorities and to engage in dialogue across disciplinary boundaries. The first Colmar Colloquium, here reported, was conceived in this spirit. Participants were asked to make explicit their assumptions regarding the relative evaluation of scientific knowledge; how in disciplines of different character one could distinguish the good from the less good; what is that identifies a significant discovery or breakthrough or other major advancement. More specifically the objective of the colloquium was

- to *promote* communication and understanding of ideas concerning advancement in different disciplines and to compare disciplines in this respect.
- to *identify*
  (a) criteria of advancement as they are actually applied;
  (b) the degree of agreement or disagreement on selection, definition and application of such criteria in disciplines of different character.
- to *examine*
  (a) whether there is any current re-evaluation of criteria because of internal or external circumstances;
  (b) the obstacles to advancement and means to promote advancement.

Obviously a meeting of this kind had to rely on the opinion of representatives of a limited selection of disciplines. The chosen fields were physics, mathematics, biology, medicine, sociology, linguistics, art history, history, economics and ecology. Care had been taken to include persons belonging to other disciplines among those participants who were not authors or discussion leaders. The function of the background papers was to provide a pool of reflexion common to all participants in advance of their coming together.

On several occasions the impact of science on society, the public image of science and the role of the scientist as expert was brought up in the discussions, but these questions could not be dealt with at length within the format of this Colloquium.

By publishing the colloquium papers *in extenso* the European Science Foundation hopes to inspire also students outside the original microcosm to engage in similar self-study and trans-disciplinary communication. The material evidently also contains messages of significance to funding agencies and to those who are engaged in matters of science policy and university education.

A colloquium of this scope would not have been possible without the generous cooperation of many individuals. First to be mentioned are the members of the organizing committee, Sir Hermann Bondi, UK, Professor S. N. Eisenstadt, Israel, Dr D. Frederichson, USA, Professor H. G. Gyllenberg, Finland, and Professor R. Vierhaus, Germany. Professor G. Sjöblom, Denmark, produced the initial document on the basis of which the guidelines for authors and discussion leaders were subsequently worked out. That the chosen theme was timely is evident from the roster of leading scientists and scholars who responded to the invitation to speak, to write and to prepare comments.

The ancient city of Colmar provided an excellent meeting place and an urban environment which greatly helped to keep the group together also during meal-times and evening hours – an important side of intellectual bridge-building

The European Science Foundation is to be congratulated on taking this initiative and carrying it through to its successful conclusion. Nevertheless the Colloquium should be seen as but the first in a series designed to help the different disciplines of science to confront and communicate in their common search for enlightenment.

Torsten Hägerstrand
Chairman of the Organizing Committee

Pushing back frontiers – or redrawing maps!

J. ZIMAN

1 SPECIALIZATION

THE life work of a scientist is to add a few bricks to the edifice of knowledge, or to fit a few more pieces into the jigsaw. These clichés express a profound truth. Progress in science, like progress in industrial manufacture, is born out of the division of labour. Adam Smith's famous account of the many distinct operations that go into the making of a pin can be paralleled by any historical account of the many distinct investigations that go into the making of a scientific discovery. A work that would be beyond the powers of any single person is easily accomplished by the co-ordinated efforts of many individuals, each performing a specialized function.

The traditional term for a scientific speciality is a discipline. But the division of labour in research is far more detailed than it is in academic instruction. In any mature scientific discipline there is a framework of established facts and theories within which specific research projects can be very precisely categorized. This is very obvious in the classification schemes used by secondary information services to retrieve scientific information from published papers. In physics, for example, this scheme is numerically ordered to four significant figures, with two further alphabetic indexes for the finest levels of discrimination (Fig. 1). There is no generally agreed nomenclature for the different levels of specialization in science. A paper such as I might myself have written when I was active in research could have been located unambiguously to within about one-tenth of a 'subfield', which is itself about one-tenth of the 'field' of 'Transport properties of condensed matter'. But this field is only about one-tenth of one of the ten subdisciplines into which physics is conventionally divided. In other words, if you were to have asked me 'What is your speciality?' I would probably have answered 'Theory of electronic transport in metals', which would suggest that my research interests and activities were limited to less than 1 per cent of all that was going on in physics.

Physics happens to have a strong theoretical framework within which it is easy to define distinct fields and subfields of research. But research work

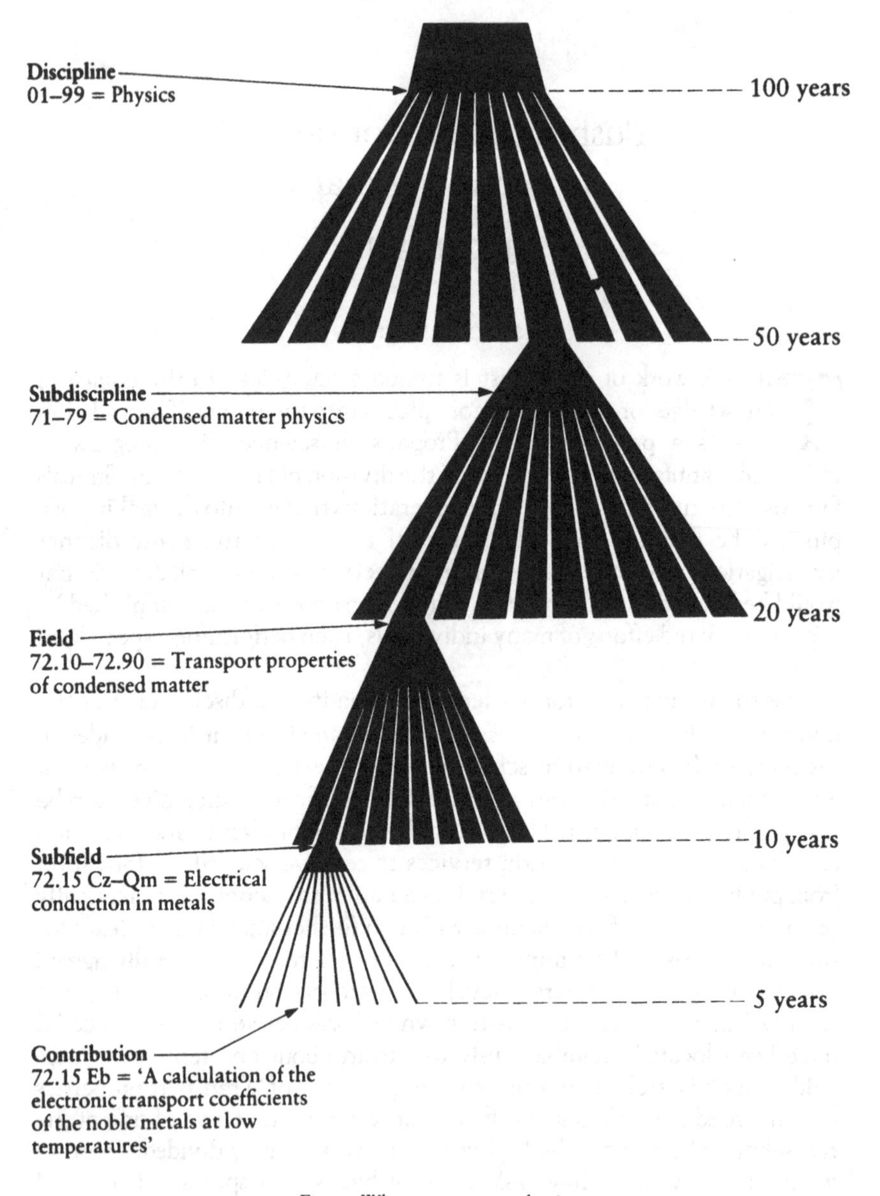
Discipline
01–99 = Physics
100 years
50 years
Subdiscipline
71–79 = Condensed matter physics
20 years
Field
72.10–72.90 = Transport properties
of condensed matter
10 years
Subfield
72.15 Cz–Qm = Electrical
conduction in metals
5 years
Contribution
72.15 Eb = 'A calculation of the
electronic transport coefficients
of the noble metals at low
temperatures'

Fig. 1. What is your speciality?

in all the other basic sciences, and in many branches of applied science, is just as finely divided into specialities. Every discipline has its classification schemes, which reflect the various ways in which research workers and research groups define their particular research interests. From the lists of publications that our academic colleagues present when they seek promotion, it is evident that this differentiation into specialities goes down to the second or third significant figure. Studies of the citation links between published scientific papers also indicate that the characteristic extent of an active research speciality is of the order of 1 per cent of a conventional academic discipline. When we say that modern science is an extraordinarily specialized activity, we mean that research is going on in several thousand more or less distinct fields, each with its particular experts, its particular techniques and its particular topics for investigation.

2 PROBLEMS

That is not to say that every scientist becomes, and remains, a narrow specialist, working in the same field throughout his or her career. Some scientists do, indeed, confine their interests within a very narrow range, but others make significant contributions over a number of fields, or even move successfully from one discipline to another. But when you ask them what they are trying to do at any particular moment, they will tell you that they are trying to solve such and such *problems*. The scientist working in an established discipline does not face a virgin natural world awaiting exploration or exploitation; he or she is normally immersed in an intellectual, technical and social milieu in which certain 'questions' seem to be demanding answers. Scientific progress is seldom achieved by boldly striking out into the unknown, or rashly speculating on the inexplicable: it arises primarily out of concentrated attacks on relatively well-posed questions that have arisen from previous research. If there is ever to be any hope of getting a reasonably convincing answer in less than a lifetime of effort, these questions must be of restricted scope. In other words, *a feasible research project can only be conceived in relation to a specific scientific problem within the bounds of a specific scientific speciality.*

One of the main tasks of the sociology of science is to understand how the labours of innumerable scientists, working away at innumerable scientific problems, are co-ordinated and integrated. One of the tasks of the philosophy of science should be to analyse the notion of a scientific problem. It cannot be just like a crossword puzzle, or mathematical exercise, since it is not known to have a unique solution. It arises out of the research process itself, and must therefore be both novel and open-ended. If there were any standard way of solving it, it would be a routine exercise, not a genuine scientific problem. And yet it must be formulated with sufficient clarity and precision that what would count as an acceptable solution can seem within reach.

The context within which a scientific problem is posed is thus just as significant as the question itself. This is what Thomas Kuhn really meant when he described 'normal' science as 'puzzle-solving' within an unquestioned 'paradigm'. In the terminology of machine intelligence, the context of a problem is its frame, defined as 'a collection of questions to be asked about a hypothetical situation; it specifies issues to be raised, and methods to be used in dealing with them'. Scientific specialities are just such frames for the formulation of potentially soluble scientific problems. Sometimes – as in the case of Darwin's problem of the origin of species – the frame is enormous, taking all natural history as its 'problem area'. In most cases, the frame of a problem is really very narrow indeed. At CERN, during the last four months, they have been spending vast sums to observe just half a dozen cases of a special type of particle collision out of the millions that are occurring in the apparatus. Of course there are very good reasons for thinking that it is important to answer the question 'Can we find empirical evidence of the hypothetical W boson?', but this project can nevertheless be classified to three significant figures: it clearly belongs in category 13.80: 'Phenomenology of hadron-induced high- and super high-energy interactions between elementary particles'. All the skills of hundreds of physicists and engineers are concentrated into that narrow salient.

3 DISCOVERY

It is one thing to formulate a good scientific problem; it is quite another matter to solve it. Only a small proportion of all the research projects that are started arrive at really significant results. Many fail to get any results at all. Others produce results that are later shown to be invalid. Of those that finally survive the successive stages of critical assessment and accreditation, the vast majority are of no great scientific interest.

Scientific progress is not made by the accumulation of routine research results: it is made by important scientific *discoveries*. This is another notion that calls for careful philosophical analysis. It certainly means more than the acquisition of information. Like the notion of a problem, a discovery cannot be defined without reference to its context. In common usage, the word carries with it an air of surprise. An investigation that arrives at exactly the result we expected can scarcely be referred to as a discovery: the information obtained must add to, or significantly change, what we already knew. In other words, it has to be considered in relation to a particular frame of existing knowledge.

Scientific discoveries, like scientific problems, have to be located in relation to scientific specialities. We automatically judge a discovery to be important or significant as soon as we see that it has implications for an extensive set of existing problems over a wide range of specialities. The other contributions to this meeting will undoubtedly show that this basic epistemological principle applies in almost all scholarly disciplines. It obviously underlies the practice of using citation counts as indicators of the quality of scientific papers and their authors. This is not, of course, a reliable index of the value of a particular contribution to knowledge, but it accords closely with the way in which all scientists think about the progress of their subjects.

But the speciality to which a particular discovery relates is not necessarily that from which it originated. As every scientist knows from experience, an investigation that was undertaken to solve problem *X* very often produces the answer to some rather different problem Y. Columbus set sail for China, and came upon America. Copernicus set out to improve the calendar, and discovered that the Earth goes round the Sun. It is not so much that many scientific discoveries are serendipitous: it is that they can have implications for quite a different set of problems than was originally foreseen.

Those philosophers who are scornful of what Kuhn called 'normal' science have tended to see its sole outcome as the accumulation of minor discoveries that have little effect outside the fields or subfields in which they were made. They tend, therefore, to concentrate on those bold hypotheses or experiments whose initial problem frame extends over a whole discipline – Wegener's theory of Continental Drift, for example, or the contemporary experiments to test non-locality and/or causality in quantum physics. But an important discovery that can eventually revolutionize 100 specialities may arise out of the investigation of a very 'unimportant' problem. Remember that quantum physics itself began in the narrow speciality of theoretical statistical mechanics, as a fudged solution to the problem of calculating the amount of light produced by a red-hot object, and that all the sciences and technologies that use the nuclear transformations of matter arose from the peculiar phenomenon of radioactivity observed in a few uncommon minerals.

4 MAPS OF KNOWLEDGE

The influence of a really important discovery is felt over substantial parts of one or more scientific disciplines. Whole subdisciplines may be revolutionized by it, as seismology and vulcanology have been, for example, by the discovery of continental drift. We tend to think of a typical scientific revolution as something that mainly affects knowledge about a particular 'subject', but the effects of a discovery are not necessarily concentrated within a standard category of the classification scheme of a discipline. The discovery of radioactive isotopes, for example, not only created the new subdiscipline of nuclear physics: it has also had an enormous effect on the experimental techniques of many quite distinct subfields of chemistry, biology and geology. If we were to mark these effects on the classification scheme of one of these disciplines, we should see them scattered here and there, apparently at random, and not concentrated in any particular region of the scheme.

There is nothing mysterious about this. For excellent practical reasons, information scientists construct one-dimensional schemes where each speciality is labelled by an alphanumeric symbol, such as 72.15 Eb. These symbols can be arranged hierarchically, or along a line, but this does not correctly represent the neighbourhood relations between the categories they stand for. Thus, for example, in Fig. 1, subfield 72.15 'Electrical conduction in metals' is properly classed next to subfield 72.20 'Electrical conduction in semiconductors', but is quite a long way from field 74 'superconductivity' to which it is equally closely related. A branching hierarchy of this kind simply cannot do justice to the real affinities between the specialities of a mature scientific discipline.

The next step, of course, is to try to represent these affinities on a two-dimensional map, as in Fig. 2. It is, indeed, possible to construct such maps, where citation linkages form coherent clusters, corresponding to recognized specialities. But although such a map may be good enough to distinguish between specialities that would otherwise overlap, it does not show all the connections that could be made between the problems that are studied in these specialities. The topology of the network of affinities between scientific specialities is really immensely complicated. A typical scientific discipline is much more compact, much more tightly interconnected, than can easily be represented on a two-dimensional diagram. On almost any subject of research, there is a dimension of theory and a dimension of fact, a dimension of methodology and a dimension of potential applicability. For the specialist there are further dimensions, or at least perspectives, associated with different schools of thought, different approaches and different research goals. This multidimensional complexity of the real 'map' of scientific knowledge is, of course, embodied in computer techniques for searching the literature and retrieving relevant information. The computer is told to find any information that matches the items in a certain list of features – for example, noble metal, conductance, electronic, low temperature, theoretical – without regard to

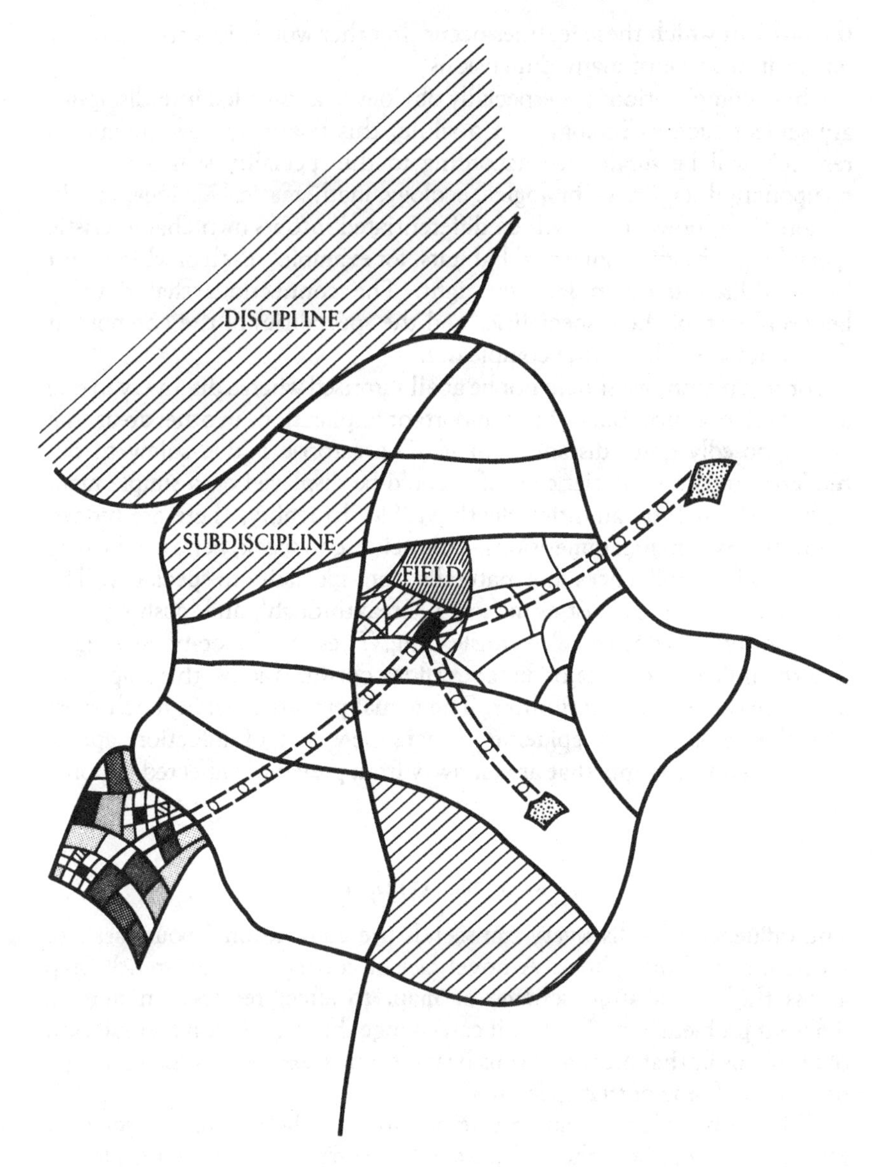

Fig. 2. The influence of a discovery on the map of science.

the order in which these features occur. In other words, it is conducting a search in a space of many dimensions.

These complications are especially obvious in a so-called interdisciplinary subject such as ecology. For a while, this relatively new domain of research will be subdivided according to the speciality structure of its component disciplines – biological ecology, mathematical ecology, etc. In a short time, however, it will be differentiated into its own characteristic specialities – by environmental habitats, for example – each of which must be multidisciplinary in its own right. The connections that develop between each of these specialities and the specialities of the component disciplines are of byzantine complexity.

For this reason, we should not be at all surprised when a discovery in one area of science turns out to have important implications in other areas that are supposedly quite distant from it. The surprise is due solely to our tendency to think of science as if it could be displayed on a simple map, with all the natural affinities clearly visible. In reality, there are hidden connections in higher dimensions – 'tunnels', so to speak, beneath the map – along which influences flow naturally from speciality to speciality. The stereotype of the scientist as making a 'breakthrough', and 'pushing back the frontiers of knowledge', is misleading. These trite metaphors suggest that the influence of a major discovery flows outwards across this map, with a continuous expanding frontier. The actual process is often much more like the spread of an epidemic, where new foci of infection appear unexpectedly at points that are far away from previously affected regions.

5 TRANSPECIALIZATION

The influence of a discovery transcends the conventional boundaries of specialities and disciplines. A novel fact or concept can apparently leap across the classification scheme or map, to affect research in quite a different problem area. That is, it can change the formulation and solution of problems in that area, and thus become a new element in the paradigm or problem frame of that speciality.

This is obviously of immense importance in the progress of science. Historians and philosophers of science have often drawn attention to the power of analogy in the work of discovery. It is sometimes argued that all intellectual progress can be ascribed to the transfer of an existing model or concept from its conventional speciality to a new problem situation, into new combinations with other facts and concepts. But this is only one form of a general phenomenon of transpecialization which also includes the transfer of empirical facts, experimental techniques and other problem-solving resources from speciality to speciality, without regard for academic frontiers.

All scientists are taught that they must not think of their discipline, subdiscipline or field of research as a closed domain, self-sufficient for its own development. But this precept is usually interpreted rather narrowly, and applied only to the immediate interspeciality neighbourhoods. Theoretical physicists, for example, are taught that they should watch for new developments in applied mathematics, and sociologists of knowledge are now labouring to keep their minds open to new trends in epistemology. But this conventional policy does not take into account the influences that can travel from great intellectual distances and prove themselves fertile on new ground. Theoretical physics, for example, occasionally gains valuable insights from certain branches of very highbrow pure mathematics, and also from experience with very practical problems in engineering.

A fundamental lesson of the history of science is that every speciality should keep itself open for such latent influences from other distant specialities. This is, of course, a counsel of perfection, since it seems to call for just those acts of imagination that are labelled 'creative'. Nevertheless, it can be fostered by certain strategies of education and research management. There is some evidence, for example, that the influence of a discovery is transferred more by outward push then by inward pull. Individuals who have been 'infected' by new ideas, techniques or facts carry them with them when they migrate into new problem areas. This would speak for a policy of encouraging intellectual mobility in the research system. It would be interesting to document cases of the other kind, where researchers resident in a particular speciality hear of some new discovery in a distant field, and realize its potential applicability to their own problem. Such events would presumably be increased by an emphasis on breadth and diversity of interests in training for research.

Modern science is moving from individual to collective modes of research. Institutes and project teams are created to study valuable natural objects, such as forests, or to develop artificial devices, such as nuclear reactors. The problems of mission-oriented research and technological development are unavoidably transpecialized, if not transdisciplinary, and can be tackled only by groups of people with diverse specialized skills and experience. The daily intellectual transactions within such a team undoubtedly encourage the transfer of new knowledge from speciality to speciality. This is not just a one-way process, from the basic sciences to their multidisciplinary applications. The problems that arise in engineering, medicine, agriculture, etc., set further tasks of fundamental comprehension for the basic natural sciences, and thus influence the paradigms of their specialities.

6 REDRAWING THE SCIENTIFIC MAP

Every scientific speciality is being transformed continually by internal discoveries or external influences. Sometimes this transformation is slow and uninteresting, as in the accumulation of data by routine investigations; at other times, an exciting discovery opens up many old problems for solution, and suggests many new ones. In the latter case, the subject may grow rapidly, both in the amount of research being done and in its intellectual scope.

But a scientific speciality cannot expand by a large factor without rupturing the classification scheme by which it has been defined. It is not enough to obtain a whole lot of new answers to old questions: if there is to be genuine progress, this information must be brought under intellectual control, and made the basis for new theories, new explanations, new concepts – and new questions. But the original definition of the speciality, and the way in which it is embedded in any larger classification scheme of the discipline, embodies current scientific opinion on just such matters. A 'map' of a discipline or field is itself a paradigmatic theory of that field, and is subject to revision as new knowledge is acquired. In other words, the notion of scientific progress is inseparable from the generation of new theoretical structures, new paradigms and new definitions of disciplines, fields and other categories of specialized knowledge.

Any major transformation of the problem frames and problem areas within a particular field of research can be described as a revolution of thought within that field. We need not enter here into the great historical and philosophical debate concerning the way in which such a revolution occurs, and whether it represents a change to a state of knowledge that is incommensurable with what it was before. What we can assert is that the way in which knowledge is conventionally mapped within that field has been significantly altered, to the extent that the older speciality frames no longer guide the researcher on problems to be solved and the means that might be used to solve them. Whether this change has been almost imperceptibly slow, or frighteningly fast, the time has come to delineate new speciality boundaries, and to draw new 'maps' of the subject.

When we are considering scientific progress in the large, we are bound to emphasize major paradigm changes that revolutionize whole disciplines. But the same process of growth and change occurs at every level of specialization, from the narrow little subfield to the major discipline. In my own former speciality of theoretical physics, for example, the subfield concerned with the electrical conductivity of liquid metals was revolutionized some 20 years ago by the discovery of a simple formula that at last made sense of all the old experimental data. I have not been following the subject recently, but I have no doubt that it has gone through several more changes of understanding and interest since that time. On a much broader scale, the modern subdiscipline called 'the physics of condensed matter' came into existence about 40 years ago, with the application of quantum theory to the behaviour of electrons in metals. But this was only one episode in the vast revolution of physics by the discovery of quantum phenomena at the beginning of this century.

As this example shows, the narrower the definition of a speciality, the shorter the time-scale on which it is likely to change. Citation analysis shows that the average 'half-life' of a primary paper in physics is about five years. If it has not made its mark by then, and been incorporated into a larger movement of change, then it will probably fall into oblivion. My guess is that the lifetime of an active 'subfield' of physics is about 10 years, and that 'fields' of research grow and decline in periods of the order of 20 years. A typical subdiscipline, such as nuclear physics, probably retains its identity for 40 or 50 years, while the transformation period of physics as a discipline is something like a century.

These time constants are, of course, purely notional. Some specialities change much more rapidly than this spectral analysis would suggest, while others continue unchanged for long periods. The main point is that the map of every science – and of the sciences as a whole – is continually being redrawn at different rates on different scales. The occasional grand revolutions that are so visible to the historians and philosophers do not cover all the radical progress that is actually being made.

7 ADAPTATION TO CHANGE

It would be interesting to look at the historical development of several different disciplines, to see whether they all redraw their maps on similar scales at similar rates. One of the fundamental differences between a natural science and a branch of humanistic scholarship may be that the latter changes much more slowly, and retains its basic problem areas and frames for much longer periods. It may be, also, that the classification scheme of a discipline is not a significant indicator of its internal structure. Such a scheme is never more than a conventional device for putting books on library shelves or finding items of information in a subject index. At any given moment its categories are only roughly representative of patterns of specialization that are continually in flux. Every active research worker has his or her own perceptions of the shapes of the specialities and frames within which research has to be planned and discoveries made.

But from this point of view we see science as a vast collective undertaking, where growth and change occur on a larger scale than any individual can hope to influence. Whatever we attempt to achieve by our own efforts, the 'moving finger' of progress will write, and move on, at its own rate, and we must harmonize our own lives and careers to it as best we can.

The structure of specialities, and their characteristic rates of change, is a subject with important implications for scientists as individuals and as members of scientific institutions. It is reasonable to assume that the discipline into which we were first educated and by whose methods we do our scientific work will outlive us in its main outlines, although we may have to move into new interdisciplinary areas as these develop. For most scientists, indeed, it is possible to maintain an active research role within a major subdiscipline, such as solid state physics, or seismology, or neurophysiology, or even econometrics, over a long research career. This can be our part of the edifice: the wall, or pillar, or buttress, or ceiling which our stones have helped to build.

But any scientist who persists unduly in a narrower field of research is at risk of being left behind as specialities change, and the active research front moves elsewhere. And there can be no security at all for a scientist attempting to specialize on a finer scale. At the subfield level, there are no really stable specialities at all. As influences flow in from neighbouring or distant fields, all is in flux. Paradigms of theory, knowledge and technique change so quickly that they cannot be treated as stable frames for the definition of problems or the means for solving them. There is little hope for a scientist who cannot set up a research programme within a somewhat broader range, and adapt intellectually and technically to the imperatives of progress and change within that range. This, surely, is an important message for graduate students, for their supervisors, and for all who manage, or are managed by, the research enterprise.

Assessment in physics

H. HAKEN

I PHYSICS AS THE PROTOTYPE OF A SCIENCE

MANY scientists consider physics as the prototype of a science, and, indeed, physics exhibits a number of features which can be regarded as highly desirable. For example, the phenomena it deals with can be described in precise terms and the relevant variables quantitatively measured; an enormous variety of phenomena may be explained by means of a few fundamental laws, which are formulated in mathematical terms and allow physicists to make quantitative predictions. It is therefore of particular interest to discuss what kind of assessment is used within the physical sciences. At the same time, the historical development of physics has taught us to exercise caution with respect to the goals which we may achieve. Though we are inclined to believe that the laws of physics are fundamental, their development has shown that even such laws may be applicable only to certain classes of phenomena. When more refined measurements are made, phenomena may appear whose explanation requires still more basic laws. Examples are the laws of classical mechanics, which at the atomic level had to be replaced by the laws of quantum mechanics. Mechanics valid for particles at comparatively small speeds had to be replaced by the theory of relativity for particles at speeds coming close to that of light.

The belief in complete predictability has been shaken twice. First, around 1925 with the advent of quantum mechanics, it became clear that at the microscopic, atomic level events cannot be predicted with absolute precision. However, at the macroscopic dimension of our daily life these 'statistical' effects are wiped out, and we can safely predict the path of a rocket sent to the Moon or to Jupiter. But it has recently become clear that even in macroscopic physics some phenomena cannot be predicted with a certain precision – which sheds new light on the limits of weather forecasting. The corresponding phenomena are called 'deterministic chaos' and are the subject of research of new branches of theoretical physics (and mathematics). Finally, we must not overlook that, as in any other science, there are certain trends or 'fashions' which play a temporary role in the development of physics. Such fashions are typical of co-operative effects in the scientific community. For example, dispersion relations in high-energy physics were in vogue in this field for a while, but today few are concerned with this approach. In spite of these remarks, physics still possesses an enormous degree of precision. Even if a single event cannot be predicted with absolute precision, it is still possible to make statistical predictions which can be checked by repeating the corresponding experiment. This possibly distinguishes physics from the historical sciences, where the course of events can never be repeated. It might be worth mentioning that events in our brain can never be totally repeated, due to memory and to learning.

On the other hand, my remarks on certain limitations to physics should caution us that we must not expect a totally objective measure for assessing achievements in physics. In order to find such measures, the 'citation index' is sometimes invoked. This index shows by which other author a certain paper has been quoted, and one might thus hope to find how much impact that paper has had on scientific development. One quickly realizes, however, that such a method is not entirely reliable. For instance, as one may discover, there are 'citation clubs', in which scientists quote only each other and omit references to any other relevant publication, if its author does not belong to that specific club. The referencing system of journals may have added to the development of 'citation clubs'.

After these preliminaries, let me list what are considered important achievements in physics. However, I wish to warn the reader that the list may not be complete.

2 THE CHARACTER OF IMPORTANT ACHIEVEMENTS IN PHYSICS

Among the important achievements are the following results.

2.1 Experimental physics

(a) Experiments which serve as *experimentum crucis* to check the validity of existing or of proposed new 'fundamental laws'. A famous example is the Michelson effect through which it became clear that the hypothesis of the ether is not tenable. In order to explain this effect (and possibly others) the theory of special relativity eventually emerged. An example where a new concept was substantiated is the effect of parity conservation, predicted by Lee and Yang and experimentally verified by Woo, Schopper and others.

(b) The discovery of effects which are not in contradiction to existing fundamental laws but are unexpected. An example is the Mössbauer effect which can be derived from quantum mechanics. Before this effect was discovered it had not been envisaged by any physicist.

(c) The substantiation of predicted effects, in particular if the theory is complicated. In such a case the fundamental laws are valid, but to derive predictions from them might be difficult as different theoretical approaches predict different effects and an *experimentum crucis* is again required.

(d) Inventions, i.e. constructions of new physical devices using existing theoretical background. Famous examples are the transistor (invented by Bardeen, Brattain and Shockley), and the laser (proposed by Schawlow and Townes, and based on Einstein's concept of stimulated emission of radiation).

2.2 Theoretical physics

(a) The formulation of new fundamental laws to explain experimental results that could not be explained by the existing laws. Examples are the theory of relativity (Einstein) or quantum theory (Heisenberg, Schrödinger and others).
(b) The formulation of new fundamental laws predicting new phenomena. An example is the law of electroweak interaction (by Glashow, Salam and Weinberg).
(c) Development of general methods to derive the observed phenomena (or to predict new ones) from the fundamental laws. In particular, a powerful theoretical or mathematical method can find widespread application. An example is the renormalization group technique by Wilson.
(d) Prediction of new effects based on the fundamental laws. An example is the pronounced change of laser light statistics which I predicted (in contradiction to other existing theories).
(e) Development of a specific concept to explain experimental findings. A famous example is the theory of superconductivity by Bardeen, Cooper and Schrieffer to explain a phenomenon which had resisted an explanation for decades.
(f) Development of unifying concepts under which a variety of phenomena can be assimilated. This point might require some more explanation, especially for scientists who are not very familiar with physics.

While it is often believed that physics has reached its ultimate goal once all its 'fundamental laws' (for example, on elementary particles) are found, a change of attitude can now be observed: namely, when dealing with complex systems, a knowledge of the fundamental laws is not sufficient (and sometimes even not necessary) for an understanding of the features of complex systems. Within synergetics it could be shown that certain concepts, such as 'order parameters' and 'slaving', which allow us to cope with transition phenomena of complex systems, may have a universal character so that these concepts acquire the character of a 'fundamental law', but in quite a different context.

2.3 Experimental and theoretical research into applications

This may be the development of new electronic devices (for example, Schottky diodes) or the realization of concepts such as miniaturization.

These developments are of great importance for our modern life (computers, telecommunication including television).

3 *A PRIORI* AND *A POSTERIORI* ASSESSMENTS IN PHYSICS

My list above might be misleading, in that it may evoke the feeling that assessment in physics is a rather simple task and can easily be performed. However, there are some difficulties and we have clearly to distinguish between the situation after a discovery has been made and before it was made.

3.1 A *posteriori* assessment

In many cases, discoveries or achievements fall into the categories listed above; clearly, not all the categories have the same weight, but I hesitate to provide different weights to the individual items on the above list. One reason is that the intellectual effort to achieve such results greatly depends on the type of problem. Another reason is that the impact of physics, technology and science may be judged from different viewpoints. Nevertheless, the following statements can be made. Usually, after a few years, the relative importance of a new fundamental discovery in physics becomes clear within the physics community and then to the public (which includes all the other branches of science). The time it takes for an achievement to be generally acknowledged depends on various circumstances, i.e. if a theory is rather abstract it might require longer to diffuse and be generally accepted. The importance of new achievements is witnessed, for instance, by the amount of work following the discovery and based on it. Other visible effects are widespread applications. For instance, the development of solid-state physics has led to the invention of the transistor and other electronic devices. The evaluation of achievements by the physics community is in many cases also mirrored by the awards of prestigious prizes, which have an interesting bandwagon effect, i.e. quite a number of scientists will resume research on a subject for which a prize has been awarded. While this might have positive effects, one must not overlook the other fact that development in the field can now be overemphasized. At any rate these remarks show the high responsibility of prize committees towards the physics community and, in addition, to society.

3.2 A *priori* assessments

We now turn to the other, still more important and difficult problem, of making *a priori* assessments, which confronts one wishing to do research or to sponsor it: what goals does he seek and what does he wish to discover? A problem may be set from the outside, by society or by a company. What are the means and methods that may be applied for its solution? In a number of cases, given a knowledge of the fundamental laws of physics or basic concepts, rather reliable guesses can be made with regard to feasibility, costs, etc. This is particularly true if a related problem has been solved before. Examples are provided by nuclear technology and space technology, where even hitherto unknown problems could be solved in a well-planned manner. Things become different and, in a way, more difficult when physicists are searching for a discovery. To elucidate the specific difficulty in this case let me adopt an extreme position which, however, lies quite on the line of information theory. The information we obtain (through a discovery) is the greater the more the discovery is *unexpected*. That means, of course, *contradictio in adjectu*, i.e. we have to predict the unpredictable or, at least, the unlikely or the unexpected. We cannot ignore the fact that a number of the most important discoveries were made by chance, but chance alone was not enough. The scientist who made the discovery was often a genius who recognized that the effect was unusual and very important. Can we control the chance? To some extent we can, and by systematic research. We cannot produce a genius, but we can train scientists to be increasingly capable of recognizing significant 'patterns'.

Another difficulty in decision-making is connected with the enormous costs often demanded by modern research. Examples for such enterprises are high-energy physics (elementary particle physics with its huge accelerators), plasma physics (especially concerned with fusion research), astrophysics (with space telescopes) and energy research on a large scale. Other enterprises, at least at first glance, seem less costly and also of great importance. Examples are microelectronics, laser physics, biophysics including new branches such as bioelectronics, physics of brain processes, turbulence, the rather recent problems of self-organization and pattern formation.

In addition to the problem of costs, other questions concern the style of physics and even the question of what is fundamental. In the tradition of western thinking the analytical method prevails. In order to understand phenomena we decompose the objects of our study into more and more refined parts. In order to understand, for instance, the physical properties of a crystal we decompose it into its molecules or atoms. Other physicists decompose the atoms into electrons and nuclei, and still other physicists decompose nuclei into protons and neutrons. Others decompose the latter particles into still more fundamental particles such as quarks and gluons, and the chain does not seem to end.

However, there is a task of equal importance, namely to show how the phenomena of nature can then be deduced from such laws and this is by no means a trivial task. In fact, here the large field of co-operative effects plays a fundamental role and it may turn out that concepts to cope with complex systems are, at least, as fundamental as the discovery of a new elementary particle. Quite obviously we are dealing here with the problem of methodology, i.e. whether an analytical approach alone is valid or whether a synthetic approach, which is prevalent in eastern thinking, is also of great value.

Now let us come to the central question. Can we develop general guidelines for scientists or governmental institutions on how to assess future lines of research? From what has been said above, it transpires that there are probably no specific guidelines possible. There are two constraints: the first is limited resources, and the second the difficulty of predicting the unpredictable. Nevertheless, we learn from history that, in spite of such difficulties, we may surely overcome them, if we admit that a wrong decision will occasionally be made. Generally speaking, it will be important to retain the adaptability of any institution to new developments and, if necessary, to reform or even close existing institutions in order to open new ones. In my opinion, broad sponsorship at a rather low financial level to trigger innovations and discovery is desirable, but such sponsorship must be linked with stringent control to determine whether, after a reasonable time (to be decided for each project), it should be continued. Projects which become promising must be sponsored at a much higher level, but it will be difficult to set up general principles on assessments here because the situation is so complex. Only boards consisting of highly experienced and outstanding scientists can decide from case to case on such projects. In my opinion, it would be very dangerous to leave final decisions to politicians, even if certain scientific groups are not successful in intrascientific decision-making. Scientists should be aware of the fact that there may be problems in decision-making which do not allow for a single solution. Each solution has its advantages and disadvantages. In such a case, deadlock can easily arise. Then it does not matter which solution is adopted, but it is important that an agreement is reached among the scientists in every case. It will be necessary to retain and to establish large international centres. Apart from their efficiency in solving important problems, there are many highly desirable side-effects such as the development of mutual understanding between young people of a very high intellectual level and from various nations. In addition, high goals have always had a great technological impact on other branches. However, it will be most important for such institutions to retain their flexibility.

In general it should be stated that the principle of self-organization of scientific research must be fully retained. Modern society may, however, require that, because of limited resources, scientific results are eventually useful to its needs. There might therefore be cases where there is a legitimate desire on the part of society to set general goals for certain branches of scientific development, but it is of the utmost importance for these goals to be only broadly defined and for enough space to remain for individual evolution and self-organization of the various branches of science.

In conclusion, I should like to comment on an epistemological issue. Presumably inspired by the revolutions of physical thought caused by the theory of relativity, quantum theory and other developments, Thomas S. Kuhn developed his ideas on 'normal science' and its change through a change of 'paradigms'. It appears to me that we have a whole hierarchy of paradigms ranging from small to big ones. In this picture, the evolution of new small paradigms may lead to the formulation of bigger ones, several bigger paradigms may trigger the evolution of a still more comprehensive one, and so on *ad infinitum*.

Comment

C. RIZZUTO

Before focussing on Professor Haken's paper, I shall touch on a few general points, and refer briefly to some of the papers whose discussion will follow this one on physics, thus hoping to reach a more unified picture.

The first point is related to the difficult problem of making everybody aware of differences in vocabulary. Differences in specific terms can be very wide between the scientific environment and the outside world and remain sensitive even in our milieu; they are sometimes increased by the use of a *lingua franca* (English, in this case) which may be transformed into a 'broken' language. As the first example of a specific word which changes its significance, let us consider the noun 'theory', or the adjective 'theoretical', which, in the laymen's vocabulary outside this room, means approximate, 'non-practical', and is, by laymen, used to describe our university or research work, both when it is formal (based on mental construct) and when it is experimental (based on hardware and experiments). A difference in the meaning of 'theory' or 'theoretical' is, however, also to be found within our disciplines. In physics these words are directly related to the mathematical simulation or prediction of phenomena, but in sociology they probably refer more to a working hypothesis, while they are probably meaningless or are yet to be used in medicine, and so on.

Similar differences are evident for other key words. For example, 'experimental' (which, to the layman, sounds 'vaguely unreliable') is probably meaningless for a mathematician, and presumably felt to be impossible or even dangerous in economics and sociology (also in astrophysics and in geophysics). Difficulties in finding a well-defined and reproducible meaning may appear in other key words such as 'instrument', 'application' and even 'science' itself. They are often damaging in everyday confrontation with the particular outside 'laymen' involved in the administration, political planning and decision-making of the various activities in which we are engaged. One of the objects of this colloquium, 'Obstacles to advancement and means to promote advancement', is strongly related to difficulties in communicating with our governments and administrators, and any suggestion or action which would lead to an improvement in this aspect would be a very positive outcome.

As a second point, I would suggest, with reference to Professor Gyllenberg's paper on biology, the use of A. Weinberg's three-level scheme of classification of the effects of research in relation to the assessment, namely:

(a) internal impact within own discipline;
(b) external interdisciplinary impact on other disciplines;
(c) impact (either cultural or applied) on society.

Coming now to Professor Haken's presentation of assessment in physics, may I summarize his points and my comments in the following sentences.

Physics may be considered as a prototype science. As a self-critical physicist, I should say that this statement may sound a bit too arrogant, and may exclude other disciplines that have similar performances in some subfields (such as chemistry). On the other hand, it may be said in general that, in physics, the interplay is clearer between theoretical prediction or modelling and experimental confirmation, and the rules governing this mutual interplay and the overall advances are reasonably clear (*e.g.*, the rule of reproducibility of experimental tests or measurements).

The evolution of physics, as seen by the succession of theories, can be described by the development of more advanced instruments which are not necessarily more precise or definite (for example, in a geometrical localization), but which, however, are able to explain indeterminacies (such as the quantum or the statistical indeterminacy).

The development efforts in physics and the related advancements evolve in an uneven manner, and various collective effects of a 'sociological' character may be recognized (for example, fashions, schools and bandwagon effects). The in-group competition is, however, a common source of the drive to advancement which supplements scientific curiosity.

Very rightly, a warning is issued on the assessment-of-advance problem: any mechanical (or parametrical) type of assessment is dangerous. Citation numbers, however they are analysed, and if they have any impact on the career or on the research granting systems, are easily manipulated by the people to be assessed. Citation clubs or other citation techniques can be developed to maximize 'profits'.

A brief examination of citation patterns in a very narrow subject with which I am familiar allows me to pinpoint a series of citation techniques which can all be misleading. They are as follows:

(a) citation of the Master (or of other suitable 'important' person): a typical technique in other fields such as economics;
(b) a citation club;
(c) citation of all your previous papers;
(d) random citation (of the papers which you happen to remember on the day of the deadline for submission);
(e) review citation (trying to show that you have been very thorough with the existing literature);
(f) staircase citation of only a few papers 'and references therein'.

The same warning is true, as pointed out by Professor Wulff in his paper, for less sophisticated assessment techniques, such as that of counting the papers, which can be easily multiplied by splitting one reasonable paper into a number of letters and conference reports. It can be said that any numerical assessment procedure introduces a change of the measured data system in much the same way that physical measurements of atomic systems change their quantum states.

In order to rationalize the problem of assessment, we can try to group some efficiency indicators which may allow a better assessment, but these must evolve with the field to which they are applied. To exemplify the results we must classify them, and a scheme such as that proposed by Professor Gyllenberg, or that proposed by Professor Haken (more related to physics, dividing the field into experimental, theoretical and applied) could be used. It should be stressed, however, that, in any case, assessment will require the expertise of people who know that field of research, and cannot be left to administrators and others with a very different basic education.

Assuming that the previous two points can be solved in a reasonable way, another very crucial choice confronts us if assessment has to be used, and that is of making decisions on funding. Should this assessment be *a priori*, *a posteriori* or contemporaneous relative to the scientific activity? While it is easy to exemplify the case of *a posteriori* assessment and while a contemporary assessment could, in principle, be made by observing the number of workshops or the intensity of excitement in some institutes, the *a priori* assessment of how a certain field in science or a certain research project is going to perform is a very sensitive problem.

This third type of assessment is, in fact, the one which policy-makers would like to master and use for the allocation of resources. To understand the problem better, it is useful to define three different classes of goals which can be proposed:

(a) Searching for an 'unexpected' discovery: this invariably requires a wide range of well-conducted and systematic 'frontier' research by researchers of excellent quality.
(b) Aiming to reach a completely new goal, on the basis of known pattern and technology (for example, discover the W particle, or reach the temperature for the superfluidity of ^{3}He). An approach of the 'space programme' type requires a well-focussed and well-managed use of both human and financial resources, and well-selected but opitimized team-work.
(c) Deciding to develop a given known field: this means strengthening and enlarging the basis of know-how and increasing the number of research groups in a co-ordinated way. An example could be that of enlarging solid-state physics oriented to applications, to allow the development of very large-scale integrated electronics (VLSI) or solar photovoltaic cells, or also to develop advanced alloys to save weight and energy. This type of progress needs well-coordinated funding and both research and technical resources over a reliable, long-lasting programme.

Given the above analysis we can now come to the rules that should be followed in funding research and in relating this activity to assessment. Only *a posteriori* assessment is really possible, and therefore the main effort must be directed to finding early enough the research people who have the best potentiality to reach new goals. This can only be done by base-level funding of a very wide range of programmes or areas to trigger as many research lines as possible and then by making regular scientific checks to detect and select the 'seedlings' so germinated and giving increased support to the successful ones. Repeating the process until a few well-developed 'centres of excellence' exist would allow a basis to be built for more extended programmes. We have thus allowed for the coexistence of *a posteriori* and *a priori* assessment in order to observe the 'rate of evolution' and decide where the best perspectives lie. This mechanism can be implemented only if it is carried on for several years.

Finally we come to the last crucial point: who assesses the performance of the 'assessors'? In other words, who can be the best at making an assessment and controlling the continuous efficacity of research activities? It is obvious to us, but should be strongly affirmed outside, that only very good scientists can do this in the proper way. The risk of this process being taken over by administrators or politicians lacking previous research experience is ever-increasing, especially as the requirement for administrators to have had good scientific experience is not made more customary (or is even opposed by the government authority). It could be pointed out that 'laymen' only seem to understand this principle fully when the coach of the local soccer team has to be selected – he has to be an ex-player, possibly good – but the obvious extension to management does not seem to hold.

This final point appears in the paper under discussion but is also underlined in those on economics and on art history, where some of the adverse effects of increased political attention to research problems are indicated. In many nations, this increased attention is also arising in fields traditionally immune from such involvement, and an effort to distinguish between the direction responding to outside goals and those of scientific assessment will be increasingly necessary.

In conclusion I will list a few more points which have not been directly touched by the paper under discussion. These are as follows: the correct management of age distribution in research (i.e. avoiding 'monochromatic' age groups and 'age gaps'); the need for a wide base of fundamental research wherever a 'reservoir' of applicable research and of know-how has to be obtained; the backlash effect, through changes in the job market, of a successful application or development on disciplines without consistent numbers of students; the problem of the low average scientific base of higher education, and the slow percolation of scientific evolution to the general public; mobility and transfer of scientific excellence, through the movement of people, and how it is decreasing today.

Identifying progress in mathematics

M. F. ATIYAH

1 INTRODUCTION

THE purpose of this symposium is to try to identify the various criteria which are used in different disciplines to measure 'progress'. Such criteria are rarely formulated in conscious and precise terms. They tend to be buried in amorphous form in the general culture of the discipline, and while there is, hopefully, some overall consensus in principle, any detailed attempt to clarify the issues is bound to be subjective and possibly controversial.

Rather than pretending to know what criteria are actually being used, consciously or subconsciously, by the mathematical community, I think it is therefore better if I try to describe what I feel these criteria should be. I will at the same time indicate the practical difficulties, the areas of disagreement and the pitfalls and dangers that we face now and in the future.

Before going further I must in all honesty admit that I write from one particular vantage point. Mathematics is a very wide discipline ranging from the verge of philosophy at one end, through the traditional areas of pure and applied mathematics to the newer applications centred around statistics and computing. It is unlikely that there would be unanimity of aim and purpose across such a wide range. The fundamental question 'why do we do mathematics?' would undoubtedly elicit many conflicting answers, and I cannot, of course, represent all these shades of opinion.

I write as a mathematician who has worked in several parts of pure mathematics, has had extensive contacts with mathematical physics and is aware, in a more nebulous way, of the wide range of applications of mathematics. I will try to present as objective and balanced an account as I possibly can.

2 SPECIAL FEATURES OF MATHEMATICS

Since this symposium is an interdisciplinary one, it may help if I begin by pointing out a number of structural ways in which mathematics differs from other disciplines both in the arts and in the sciences. These differences have various implications: some produce special problems and difficulties, while others are a positive advantage.

The first singular feature of mathematics is that it is very difficult to describe or define its subject-matter or content. Physics is the study of the physical world, biology deals with life, history is about mankind's past, but what is mathematics? This question is not one of mere sophistry; it indicates a genuine uncertainty about the nature and purpose of mathematics which reflects itself in quite different philosophical attitudes and value-judgements. These deep-seated differences, which I shall return to later, are, however, usually latent or subliminal and, at the more detailed technical level, mathematicians are rarely in serious disagreement.

The second and rather familiar characteristic of mathematics is its perfect logical progression. Greek mathematics remains as valid today as it was 2000 years ago, while the calculus of Newton and Leibnitz has survived 300 years without essential change. In both cases we now understand the foundations somewhat better, but their essential truth was never in serious doubt. By contrast, theories in natural science have a less certain future: they may be discarded completely like the theory of 'Phlogiston' or the notion of a flat Earth, or they may be modified and superseded in the way Einstein's theory of gravitation has replaced that of Newton.

This 'finality' of mathematics is, in its way, the envy of other disciplines which will frequently borrow mathematical ideas and techniques to lend greater weight to their conclusions. Gratifying though it may be to us mathematicians to see our ideas being put to fruitful use in other fields, there is always the danger that the logical prestige of mathematics is being used to bolster unsound arguments. The truth is, of course, that the certainty of mathematics is directly related to its separation from the real world. As soon as it is applied to any realistic situation, whether in the physical or social sciences, the conclusions share all the uncertainties associated with the experimental data or the scientific hypotheses. Any exaggerated claims about the applications of mathematics, based on the doctrine of mathematical infallibility, are likely to have damaging repercussions. The dangers are all the greater because, for most people, mathematics is a 'black art' shrouded in mystery.

The last peculiarity of mathematics is its independence. Unlike the experimental sciences, mathematics has no need for expensive equipment: it is a cheap science. All that is needed is paper and pencil, and even this is dispensable: Archimedes drew figures in the sand! Many of the great mathematicians of the past produced their masterpieces in very difficult circumstances, while in more recent times we have the example of Jean Leray who revolutionized modern topology while in a prisoner-of-war camp. Even by comparison with arts subjects the requirements of mathematics are modest: large voluminous libraries with ancient manuscripts are not needed. Moreover, unlike the social sciences, mathematics is largely independent of the political and social system. It has flourished, and continues to flourish, under political régimes of all complexions.

This (comparative) independence of financial and political factors has its drawbacks and dangers. It can lead to the total intellectual isolation of mathematics. In fact the picture I have drawn is much too black and white: mathematics through its various applications is not divorced from the real world, its rate of progress is still dependent on government funding, computers are expensive and libraries play a vital role. Nevertheless, I believe the relation of mathematics to society is, in many respects, unique and I shall discuss this relation in more detail in the final sections.

3 THE ROLE OF PROBLEMS

I have referred earlier to the difficulty of answering the question 'what is mathematics?' One possible answer is that mathematics is a collection of ideas and intellectual techniques for solving 'problems'. This may appear unsatisfactory since it begs the question 'what kinds of problem?' However, the essence of mathematics is that the raw material of its problems can arise in almost any field: it is not the content but the form that is important. In any case, whether one regards this as a convincing answer or not, it is undeniable that the solution of 'problems' has always played a fundamental role in the history of mathematics. I should emphasize that I am now using the word 'problem' in a quite narrow sense (for example, solve the following equation . . .) and not in terms of grand strategy. I will illustrate this with a number of examples.

My first example is somewhat hypothetical and concerns the famous theorem of Pythagoras, relating the hypotenuse z of a right-angled triangle to the other two sides x, y by the formula $x^2 + y^2 = z^2$. Presumably, at some earlier time, this would have figured as a problem: what is the formula for z in terms of x and y? The importance of this problem for the further development of geometry would have been easy to see. It would therefore have represented a major obstacle to be overcome, and many mathematical problems fall into this class.

An entirely different type of problem (and this time a genuine historical one) is illustrated by the famous 'last theorem' of Fermat. This begins with the observation that the Pythagorean equation above has integer solutions (such as 3, 4, 5), but goes on to assert that the more general equation $x^n + y^n = z^n$ never has positive integer solutions (for any integer power n greater than 2). Unlike the case $n = 2$, these other equations have no geometrical interpretation: moreover, the requirement to have integer solutions (which makes this a problem of number theory) alters the character of the problem. It is not at all clear *a priori* that this problem of Fermat is of much importance. In fact it has had a profound influence on the development of mathematics. Fermat claimed to have a proof but did not have space to write it down! Over the past 300 years many of the world's best mathematicians, fascinated by the difficulty of this apparently simple problem, have struggled with only partial success to prove Fermat's 'theorem'. In the course of their efforts many new techniques and concepts have been introduced and these have permeated large parts of mathematics.

Fermat's problem, then, has played the role that Mount Everest played for mountaineers (before its successful ascent). It was a challenge, and attempts to climb it stimulated the development and perfection of new skills and techniques.

As my final example I would like to discuss another famous problem: the four-colour problem, whose (positive) solution asserts that only four colours are needed for any map on the globe if adjacent countries are always required to be of different colours. This problem resisted solution for about 100 years and was finally solved, in recent times, by an extensive use of computers. However, the impact on mathematics has been small. The problem itself was not a fundamental one and attempts to solve it have not generated a body of techniques that are of great significance. It appears that it will go down to history as a curiosity, famous (or infamous) as the first non-trivial problem in mathematics solved by a computer. Of course, the situation would change if tomorrow some young mathematician were to develop a brilliant theory which *inter alia* solved the four-colour problem.

These examples bring out several points. A problem may in itself be of fundamental importance, representing an irreducible obstacle to further progress. In that case, any solution represents progress and is gratefully accepted. In many cases, however, it is not possible in advance to predict how important a particular problem will turn out to be. If it is solved rapidly by standard methods it loses much of its interest. If it resists all known methods for a long time, and enters the list of classic problems, it acquires a potential interest as a challenge. But, as the four-colour problem shows, even this provides no guarantee against an anti-climax. The real criterion for a 'good' problem is that the search for its solution should generate new powerful techniques which have a wide range of applicability. Fermat's last theorem is the standard example of a good problem in this sense.

At any given time mathematics has a large supply of problems of many types, and steps towards their solution, particularly steps involving essentially new ideas, provide one of the main indicators of progress. This is universally recognized, since all mathematicians, whatever their speciality, are essentially craftsmen and appreciate the technical skills involved in ths solution of long-standing problems.

4 INNOVATION

There is little doubt that innovation is vital in the progress of mathematics and it usually ranks at the top of the scale of criteria. This is not surprising since mathematics is almost totally theoretical and lacks the strong empirical basis of the other sciences. Mathematical progress is not stimulated by experimental work, by the introduction of technology or by the discovery of forgotten manuscripts. Progress has to be generated internally.

Innovation can take many forms. The most common is the invention of new techniques for the solution of problems. The degree of innovation here can, of course, vary from those small steps which all workers make almost daily to the giant steps which involve radically new methods. These more radical changes frequently involve the introduction of totally new concepts and require a complete alteration in earlier points of view. A classic example of this is the work of Galois on the insolubility (by square roots, cube roots, etc.) of the general polynomial equation of degree at least five. Whereas quadratic equations and subsequently those of degree three and four had been successfully solved, mathematicians had got firmly stuck at degree five. Galois showed that the problem was insoluble, realizing that the key point lay in the symmetries of the five solutions of the equation. He thereby laid the foundation for a general theory of symmetry (the theory of groups) which is one of the most profound and far-reaching of all mathematical concepts.

Radical innovation of this type usually arises in the attempt to solve a hard problem. There is, however, another form of innovation which is just as vital and that is the formulation of new and important problems. As I indicated earlier, the importance of a problem is not easy to assess in advance of its solution, so that the intelligent choice of problem requires great insight. Sometimes problems will present themselves naturally in the course of an investigation: the internal structure and cohesion of the theory essentially forces the mathematician to pose the questions. In other cases the problems may come from outside mathematics, from some neighbouring scientific discipline and I will have more to say on this later.

In general one can say that mathematics progresses by a continuous application of standard methods interspersed with spectacular breakthroughs when new concepts and problems suddenly appear. The rate of progress in any given branch of mathematics is heavily dependent on the frequency of such breakthroughs. For this reason the centre of excitement in mathematics can shift quite rapidly and unpredictably from area to area. For example, three-dimensional geometry, which was comparatively dormant until a few years ago, has suddenly come to the forefront because of the very remarkable discoveries of William Thurston of Princeton. In four dimensions a very recent breakthrough by Simon Donaldson of Oxford has attracted widespread attention because he solves a long-standing problem by entirely new methods having their origin in theoretical physics. By the criteria I have been indicating this would score highly on all grounds, and it is likely to open up a new field.

In a subject as well-organized and heavily structured as mathematics there are plenty of sign-posts and well-lit roads to guide the traveller. However, journeys along long, straight roads are regarded as dull and mathematicians lay great store by the unexpected turn. To say that something is a 'surprising' result is to bestow significant praise. When the unexpected turns up we realize the limitations of our previous understanding and we then search deeper for the explanation.

One form of surprise which is always particularly striking is the 'counter-example'. As its name suggests, this is an example constructed specifically to contradict some previously held belief. Counter-examples may be purely negative: they may indicate that no further progress is possible in a certain direction. This is in itself valuable since much effort can be wasted on the search for a 'North-West passage' that is not there. More frequently counter-examples serve merely as a warning on the limitations of various methods and act as lighthouses for the enterprising sailor.

5 THE AESTHETIC COMPONENT

For its practitioners mathematics is both art and science: beauty and truth are held in equal esteem. To the outsider, the concept of beauty in mathematics is perhaps hard to conceive and so it is worth while attempting to provide an explanation.

Most mathematicians, but particularly those of the 'pure' variety, far removed from applications, have a clear idea of what is meant by a 'beautiful argument'. This is a term of great praise indicating elegance of style, economy of effort, clarity of thought, perfection of detail and balance of form adding up to give a feeling of overwhelming conviction. Naturally, these heights are rarely reached in full, and they represent goals to be striven for, but they exercise a very strong influence. Mathematicians will frequently feel attracted to one area rather than another because they find it more beautiful, they will look for methods which are elegant and conversely they will try to avoid arguments which are clumsy or ugly.

It would be difficult to overestimate the subjective importance of these aesthetic criteria in the mind of the working mathematician. They provide much of the internal drive which propels him forward and they colour his views of other people's work. From the outside it might be asked whether mathematicians are right to attach such weight to these factors. Is mathematics not a science? Are there not more objective criteria? To some extent, particularly in the more applied areas, where mathematics merges into the other sciences, the answer is, of course, yes. But in large parts of mathematics the question is much more complex, and I would like to explain why aesthetic criteria have a proper role to play.

One of the main features of mathematics is its universality. Almost any branch of knowledge has aspects which can be analysed mathematically. The first step in such an analysis is to focus on certain topics, strip away all irrelevant material and convert what is left into an appropriate mathematical form. The success of such an enterprise will depend on the availability of appropriate concepts and formalism in mathematics and subsequently on the availability of effective techniques of analysis and computation. The development of an abstract language, flexible and powerful enough to serve a multitude of possible purposes, is therefore the essential feature of mathematics. In such an abstract world the requirements of simplicity and elegance acquire an overriding importance. To illustrate this with an ancient and familiar example, we have only to reflect on the enormous advantage of our present decimal system over the cumbersome Roman numerals. The mere idea of carrying out a long multiplication sum in Roman numerals brings home in a dramatic way the simplicity, power and beauty of the decimal system. Again the introduction of algebraic symbolism by the Arabs was a momentous but exceptionally simple step forward.

The point I am trying to make is that the development of simple and concise arguments is absolutely indispensable for progress in mathematics. It may perhaps be helpful in this respect to compare mathematics with architecture, another field that straddles the arts/science frontier. In architecture there is also the dichotomy between function and form, and while this is a perennial source of legitimate controversy, most people would agree that the best architecture arises from a harmonious marriage of the two.

I referred earlier to the difficulty of predicting in advance how significant a particular problem might turn out to be. The selection and formulation of problems is to that extent an art depending on the intuition of the individual mathematician. Aesthetic questions undoubtedly play a large role at this intuitive level.

6 INTEGRATION VERSUS FRAGMENTATION

Because of the permanent validity of mathematical truth the cumulative aspect of our subject is particularly marked. Every century that passes adds layer upon layer to the mathematical edifice, and while subsequent generations may lose interest in the fine detail nothing is totally discarded. This presents a formidable problem at the present time and an even more daunting one as we look into the future. How can this mountain of knowledge be kept under any sort of control? Will it collapse and disintegrate under its own weight?

It is clear that one effect of this large accumulation is to produce specialization. In the days of Newton, or even Gauss, mathematicians were also natural scientists. By the early twentieth century only a handful of great men such as Hilbert, Poincaré and Weyl could claim to cover even most of mathematics. Since then specialization has proceeded apace, and each generation has a narrower and narrower focus.

This process of fragmentation has been accentuated by the ever-widening field of applications of mathematics. Whole new branches come into being in response to requirements from outside, such as information theory, control theory or epidemiology. The large increase in the total number of active mathematicians has in itself also helped to encourage specialization: scientific communities need to be of a certain critical size before specialisms can exist as viable units.

Some degree of specialization is no doubt inevitable, and perhaps desirable, but carried to excessive lengths it can be disastrous. This is particularly so in mathematics whose main *raison d'être* is its capacity to transpose ideas from one field to another by the process of abstraction. Moreover, the ultimate justification for doing mathematics is intimately related with its overall unity. If we grant that, on purely utilitarian grounds, mathematics justifies itself by some of its applications, then the whole of mathematics acquires a rationale provided it remains a connected whole. Any part that drifts away from the main body of the field has then to justify itself in a more direct fashion.

The main counterbalancing force, tending to maintain the cohesion and unity of mathematics, is the development of more sophisticated and abstract concepts. At their best, these help to produce an overall synthesis, so that a multitude of special facts appear as different cases of some grand principle. In many areas this has been very successful and large parts of nineteenth-century mathematics have been absorbed, without drastic loss, in the more abstract and elevated viewpoint of the twentieth century. This explains the present dominance of a few key subjects such as group theory (study of symmetry), topology (study of continuity) and probability (study of random events).

New concepts, which help to unify past work and clear the way for further development, are therefore an essential component of mathematical progress. In the long run, they are every bit as important as the solution of hard problems or the development of new techniques. In practice, really fruitful concepts tend to emerge over a long period, in conjunction with more concrete work. Only occasionally are they suddenly created.

Fortunately, there are several other factors which from time to time help to integrate mathematics. New breakthroughs may appear which run completely across traditional frontiers. A very spectacular example of this in recent years concerns 'solitons'. These are a very special class of solutions of some non-linear differential equations, and they have many remarkable properties. They arise in many branches of physics and mechanics and so have attracted widespread attention. In the past 20 years theoretical work on 'solitons' has interacted with almost all parts of mathematics in very surprising ways, and it has had a strong unifying effect.

This example illustrates the more general fact that problems arising from outside mathematics do not always fit neatly into existing specialist divisions. Interaction between mathematics and other disciplines can therefore help, at times, to prevent fragmentation.

7 APPLIED MATHEMATICS

So far I have been writing mainly about pure mathematics, with an occasional reference here and there to outside applications. Now I must redress the balance. Mathematics has, as we all know, an enormous range of applications and it provides an indispensable language and framework for all the physical sciences. In the last analysis this must be the primary justification for doing mathematics: it is the reason why mathematics is not just the occupation of a few esoteric individuals but is a basic component of the educational and social scene.

This does not mean, however, that only mathematics of the applicable kind can be justified. As in all scientific fields, short-term emphasis on the immediate practical problems has to be balanced with long-term strategy in which fundamental pure research is pursued. Pure mathematics is not, in this respect, inherently different from pure research in other sciences. It may seem that it is excessively pure and that much of it is very far removed from all applications, but this is an inevitable consequence of the sheer variety of these applications.

I discussed earlier the pivotal role played by 'problems' in the development of mathematics. These can be either internal problems generated within mathematics itself or external, arising from some other field. These external problems provide a constant additional stimulus to mathematics and, in the long run, are essential for its vitality. Sometimes these problems fit into an existing mathematical framework and the task is the technical one of finding the right tools to produce a solution. Frequently, however, a new mathematical framework has to be created, in which the basic concepts reflect the phenomena being studied in the real world. Mathematics therefore grows both in depth and in breadth through its interaction with other fields.

The applications of mathematics are of rather different types and levels in different fields and it may be helpful briefly to comment on these in turn.

7.1 Physical science

The relations between mathematics and physics are extremely deep and go back many centuries. Large parts of mathematics, including calculus, were developed primarily in connection with mechanics and physics. Conversely, present-day physics is employing some of the most abstract pure mathematics. In general, mathematics has proved a remarkably successful and appropriate key for the study of physics and engineering.

7.2 Biological science

Applications here include such areas as genetics and population growth in addition to those arising from the physico-chemical basis of biology. Newer ideas involve problems of morphogenesis or geometric properties of large molecules, but it is not yet clear how important mathematics will turn out to be for the future of biology. Will biology generate its own brands of mathematics as physics has done in the past? One should retain an open mind and, while innovation should be encouraged, some healthy scepticism is probably wise. Excessive claims for mathematics in these new fields can easily boomerang.

7.3 Social science

The last 30 years have seen an ever-increasing use of statistics, operational research and related topics in the social sciences. In many respects, including the financial and human resources employed, this is now comparable with the role of mathematics in physics. Because this is of much more recent origin, the links with mathematics are as yet at a much shallower level. This presents some possible dangers since financial and economic pressures may push mathematics in directions which are not intrinsically as rich as its more traditional fields.

7.4 Computer science

The relation between mathematics and computer science is a very ambivalent one. Mathematicians such as Turing and von Neumann were prominent in pioneering the early developments of computers, and there is a significant use of logic in the area of computer languages. Moreover, computers have had an enormous impact in terms of producing numerical solutions to complex systems of equations. On the other hand, computer science has long since outgrown its mathematical background. The computer revolution has acquired such dimensions that mathematics is in serious danger of becoming swamped by it. This has not happened yet but I believe that it represents the greatest potential challenge to mathematics in the coming decades.

8 RELATIONS WITH SOCIETY

Of all the scientific disciplines mathematics is certainly the furthest removed from the man in the street, who has absolutely no idea what mathematical research consists of. For the layman, mathematics is identified with the elementary ideas he tried (often unsuccessfully) to absorb at school: arithmetic, geometry, algebra and perhaps the rudiments of calculus. These appear to him cut and dried subjects, devoid of all life. The idea that they were actually created in the past by human endeavour hardly crosses his mind, and so he cannot conceive that similar creative work is going on at the present time.

Of course the layman will realize that mathematics is a 'useful' subject, and that it is used by engineers, statisticians and other scientists. The idea that mathematics, as an abstract intellectual discipline, has a flourishing independent existence is, however, very difficult for the outsider to grasp. Mathematicians tend therefore to be regarded with a mixture of awe and puzzlement.

The gap between the mathematician and the layman poses very serious problems. It is difficult, though not impossible, to try to bridge it by giving popular accounts of suitable parts of contemporary mathematics. At best this can constitute only a very partial solution. The more fundamental approach, in my view, is through a strengthening of the links between mathematics and all the other disciplines in which mathematics plays an important part. Fellow scientists are in a much better position to appreciate the true nature of mathematics. Although they will only be directly involved with those parts which have a bearing on their particular discipline, this sample will provide a benchmark by which they can assess the value of mathematical ideas.

Only the support of other scientists (in the wide sense) can, in the long run, ensure that the relation between mathematics and society remains a healthy one. Even if pure mathematicians remain doubtful of the intellectual gains arising from the interaction with applied fields (and I have argued that these gains are genuine) they will be forced, out of self-interest, to look more favourably on their applied colleagues. Such a trend is already visible. After the heady expansion of the 1960s the greater economic realism of the present time is producing an increasing emphasis on the more applied areas of mathematics. I believe that this is a healthy corrective to the earlier period of unbridled purity, but equally there are dangers in an excessive lurch in the opposite direction. A situation in which mathematics becomes merely a service subject, with a few tame mathematicians attached to different research groups, would be a recipe for fossilization. An appropriate balance is essential if mathematics is to retain its integrity.

I alluded earlier to the fact that mathematics is a cheap subject and that it can progress (though perhaps more slowly) without large-scale government funding. This applies, however, only so long as the potentially creative mathematicians are not seduced into alternative careers by financial and social pressures. Newton was a great scientist but his more practical contributions as Master of the Royal Mint were hardly of the same order. The growth of computer science and the very large funds it increasingly attracts are likely to pose a danger for mathematics, not just because mathematics might be starved of funds, but because the potential Newtons of the future might be drawn away from mathematics. It may be possible to do without money but not without brains!

Although society at large may be unaware of what goes on at the research level in mathematics, it is very much concerned with mathematics as a major component of education. Society does require a large supply of numerate citizens and at this level mathematics commands general respect. By maintaining a strong and constructive interest in education and emphasizing the organic links between teaching and research, mathematicians can help to improve relations with the general public and to decrease their isolation.

9 DISAGREEMENT

Because most mathematics is presented in a formal style with 'proofs', there is rarely any serious dispute as to whether it is 'right' or 'wrong'. Of course, mistakes do occur and are not always recognized immediately, while there are occasions when a sketchy proof fails to convince the mathematical community. These are, however, comparatively rare and the passage of a few years usually clears up the misunderstandings.

At the applied end, where heuristic arguments are employed, disputes are more common. Typically, the argument will centre around the validity of some simplification or approximation, which reduces a complex real-life situation to a tractable mathematical one. Here the ultimate test is an empirical one: does the mathematics agree with the real world?

Serious disagreements are not over technical detail but over value or significance. Again, within a given field, common standards exist and value is measured by such criteria as difficulty, originality, scope, elegance and unifying power. Real disagreements begin only when comparisons are made between the relative value of different fields. How does one compare the importance of some branch of algebra with some branch of analysis? There are those who hold that it is pointless (and even dangerous) to attempt such comparative value-judgements. Let everyone follow their own guiding star, so the argument goes, and the best truth will ultimately prevail. This is the policy of *laissez faire* or, if you prefer, 'academic freedom'.

There is much to be said for this point of view and it is of course true that the value of any piece of research (in any field) is ultimately decided by posterity. Unfortunately, in the real world, decisions frequently have to be made which involve value-judgements: which paper to publish, which research student gets the Fellowship or who becomes Professor? One possible solution is always to give extra weight to those research topics which can be seen to have some direct relevance to the outside world. This would be an understandable and coherent policy but it would put at risk those long-term developments whose applications lie far in the future, and history is littered with such examples.

The alternative view, which I have been advocating, is to assess value in terms of the impact on mathematics as a whole. This is not an easy task, since one has frequently to guess what the impact is likely to be, but at least it does define a criterion. Moreover, this criterion will by its nature tend to reinforce the unifying features of mathematics and prevent its fragmentation. I believe that, in practice, great weight is attached to work that impinges on several areas of mathematics. This can be understood in purely political terms as part of the democratic process: given a committee of specialists they are more likely to support a piece of research that appears relevant to several members of the committee.

10 CONCLUSIONS

As I have tried to indicate, the criteria of progress in mathematics are subtle and complex, and are in many ways unique to the subject. It is rather ironical that a discipline which rightly prides itself on the precision of its thought and the accuracy of its conclusions should also have the greatest difficulty in defining its own criteria of value. Perhaps this is another manifestation of the Heisenberg uncertainty principle!

In looking back over the list of criteria that I have enumerated I feel that I have not perhaps given enough emphasis to the over-riding importance of quality. This is best illustrated by the case of Bernhard Riemann whose collected works occupy one modest volume but who was probably one of the most influential mathematicians of all time. Many of his papers opened up quite new fields which are still being vigorously exploited 100 years after his death. The most famous one laid the foundations of higher-dimensional differential geometry and provided the essential framework for Einstein's theory of general relativity.

Perhaps I should conclude by emphasizing that mathematics at the present time is, despite its antiquity, in a very healthy and active state. Old problems are being solved regularly and new vistas are constantly being opened up. There is little uncertainty or lack of confidence in the mathematical community. Most of my colleagues are too busy proving theorems to indulge in the kind of soul-searching that I have undertaken here.

Comment

R. THOM

Professor Atiyah's report on assessment of results in mathematics seems to me excellent down to the very last detail. My only disagreement with it concerns minor nuances. I will add some personal remarks of a general nature connected with the aim of the colloquium; these remarks seem to me to be useful in relocating mathematics amongst the sciences and in explaining the relatively favourable status of our discipline as far as its assessment techniques are concerned.

I COMMENTS ON PROFESSOR ATIYAH'S PAPER

First, I do not agree that the subject-matter of mathematics is difficult to define. Given that the aim of mathematics is to study number and space, the core of the subject has, in my view, been delineated. Of course it is the case that the possibility of running into difficulties when trying to refine this definition does exist, but as it stands it covers the main classification of mathematics in algebra–geometry/topology–analysis, as follows:

Direct generativity of operators → algebra
Geometric continuum → geometry–topology
Synthesis of operators on spaces → analysis

Secondly, I would not tend to share Professor Atiyah's view of the importance of the concept of problem and of problem-solving, or of aesthetics. The essential part of contemporary mathematical production no doubt meets this definition of problem-solving. It is the stage of puzzle-solving characteristic of the normal evolution of science as described by T. S. Kuhn. The concept of paradigm, however, is not really relevant in mathematics, but is replaced by a few major concepts, such as number, function and probability, which build up from axiomatic to axiomatic and never really reach a final form. From this point of view I would suggest that the problem of assessment in mathematics should be subordinated to a typology of results in mathematics.

2 TYPOLOGY OF RESULTS IN MATHEMATICS

Three major categories of results, plus an exceptional set, can be distinguished as follows:

(a) Results of a routine nature. These cover the elaboration of existing structures applied to situations which are more difficult and complex than before. It can be said that about 85 per cent of all contemporary production in mathematics comes under this heading.
(b) Results arising from questions posed by applied disciplines (exogenous motivation): about 10 per cent (or less) of output.
(c) Essential and profound findings, opening up new research areas. The importance of these results, which may not look very impressive as such, lies in the new perspectives that they open, the new *problématiques* to which they give rise (4 per cent of output, at the most).

These results are often linked with the major concepts mentioned in section 1 above and are therefore of endogenous origin.

Finally, an exceptional group: the solving of historical problems (or decisive progress towards solving them), such as Fermat's theorem, the conjecture of Riemann, the conjecture of Poincaré (on simply connected manifolds of dimension three) – between 0 per cent and 1 per cent of output.

This last type of result is an immediate source of fame for its author, which does not mean – as rightly underlined by Professor Atiyah in relation to the four-colour problem – that it necessarily leads to major developments, or to the opening up of vast new domains of mathematical investigation. This typology relies on a very general characterization of the evolution of mathematical theories.

Theories in mathematics have a history which can be outlined as follows:

(a) A phase of original chaos, during which a few pioneers with a germinal way of thinking are working out the essential concepts and the main problems (results of category (c) above).
(b) A phase of explosion, during which the major results are discovered, as well as the generating mechanisms and the efficient methodologies in the field under study. This is the booming period for theory-building, giving rise to an abundance of epigones.
(c) A phase of exploitation, during which, with the passage of time, problems become more and more difficult and less and less interesting.

Most results in contemporary mathematical literature issue, of course, from the third phase, and are not expected to lead to applications. From a pragmatic point of view, about 90 per cent of contemporary mathematics is useless and will most probably remain so. One should not forget that mathematical research plays the role of social selection. The identification of people able to practise mathematics is a major requirement of any society, as a fundamental part of scientific knowledge demands an understanding of mathematics – not forgetting 'deeds achieved in honour of the human mind', to quote G. Hardy's celebrated phrase.

3 COMPARISON BETWEEN MATHEMATICS AND OTHER SCIENTIFIC DISCIPLINES

The above typology could also apply to other disciplines. In all sciences, results are either endogenously or exogenously motivated (the latter in response to requests from other sciences or to technological requirements). Endogenous results are necessarily connected to the internal deductive power of theorization applied within the discipline. Thus the theory determines hypotheses, questions and experiences. Where deduction is almost non-existent or short-lived (as is the case in biochemistry and biology), endogenous motivation is linked to a local *problématique*, local in both senses of the word, i.e. as the study of a particular phenomenon, and because connected with a paradigm used locally by the scientific community. This is the reason why, in disciplines with a low level of theorization, the assessment of results is also made 'locally', or sociologically (as noted with remarkable lucidity by Professor Gyllenberg in his presentation). Using a planar diagram with an Ox-axis representing Comte's hierarchy of sciences (mathematics, physics, chemistry, biology, humanities), it becomes clear that the theoretical power which is maximal in mathematics decreases steadily but catches up again a little in the humanities as conceptual theorization. On the other hand, technological

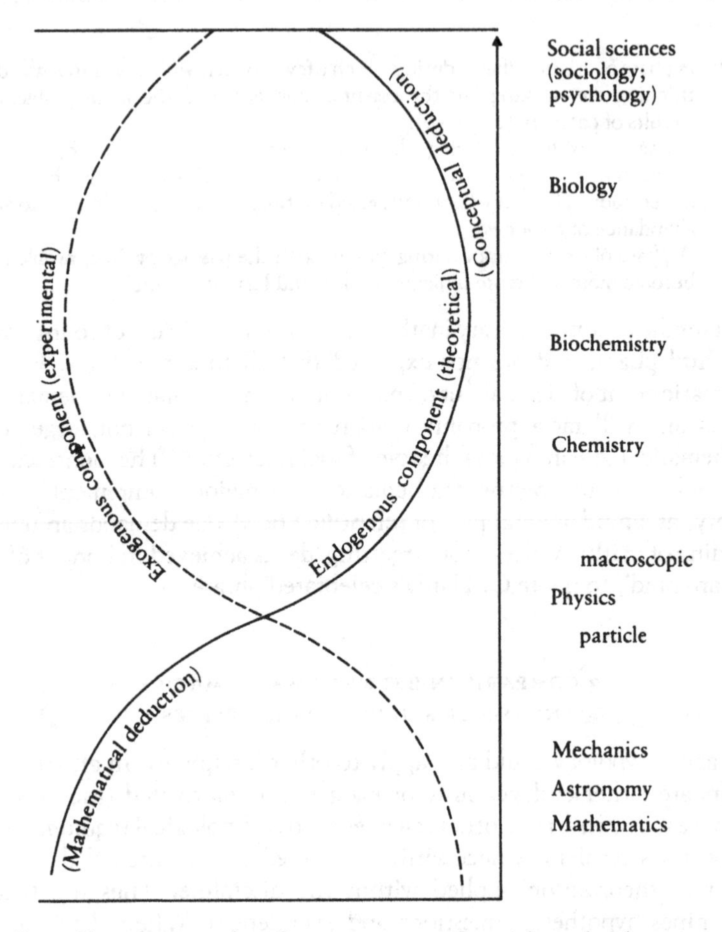

Fig. 1. Planar diagram with an Ox-axis representing Comte's hierarchy of sciences.

requirements (application) are highest in biology (where the weight of medicine – the technology of biology – is considerable). It is weak in mathematics, increases in physics, even more in chemistry and biology and then decreases again (see Fig. 1).

These considerations are important in the assessment of results. It is difficult to compare results connected to research of differing types, i.e. endogenous and exogenous. It is easier to compare two results of the same nature. If the results are connected with applications (exogenous–exogenous), the actual profit to be drawn from each result can be compared, in some cases numerically (i.e. in terms of costs). In comparing endogenous–endogenous, specialists must have a comprehensive view of their discipline in order to assess the deductive perspectives of each alternative. Such is the case in mathematics, where mathematicians in general have a global view of their discipline, even if they do not know about all recent developments. It is not the case in biology, and this explains the sociological bias to which Professor Gyllenberg draws attention in that discipline. It is a matter of fact that competition and chairs for other posts is rarely as fierce in mathematics as in other less theoretical and more experimental disciplines. This relatively fortunate situation for mathematicians may be threatened in the near future by the sociological impact of the computer and associated technology.

Progress and expectations in biology

H. G. GYLLENBERG

1 INTRODUCTION

THE last half century has witnessed tremendous development in biology (or in the life sciences, whichever concept seems to be more popular). The structure of deoxyribonucleic acid (DNA); the cracking of the genetic code; the understanding of protein synthesis; 'artificial' recombination of DNA; the description of structure and function of cellular membranes, as well as increased knowledge concerning biological regulatory mechanisms from hormones to enzyme systems are indications of obvious and far-reaching progress in the domain of biology.

This does not mean, however, that biology has solved all its problems. Sheldrake (1981) presents a list of 'unsolved problems in biology'. Although Sheldrake seems to advocate the view that a 'mechanistic' physico-chemical approach to biology does not solve problems related to behaviour or morphogenesis, this is obviously true when considering the branches of biology where breakthroughs comparable to cracking the genetic code have not yet been achieved.

The purpose of this paper is to discuss indications of progress in biology. Why do I consider cracking the genetic code as progress, whereas I feel that research on behaviour or morphogenesis of living organisms has not been equally progressive. My evaluation is certainly a subjective one, and it is probably biased by the fact that I am a microbiologist, not a neurophysiologist. This bias may also occur in the following analysis which is aimed at the evaluation of both progress and expectations in biology.

In the 1970s UNESCO conducted a study on the effectiveness of research groups in six European countries (*Scientific Productivity*, ed. F. M. Andrews, Cambridge University Press, 1979). 'Effectiveness' or 'performance' are not of course synonymous with progress, but it seems that some of the measures of performance defined in the UNESCO study, such as 'innovativeness', 'effectiveness for non R & D objectives' and 'social value of work', might be useful criteria for assessing progress. The problem is how to measure such progress. In the UNESCO study performance criteria such as 'innovativeness' were given ratings both by the researchers involved and by external experts. In addition, the UNESCO study applied 'counts of output', i.e. number of original articles published, number of patent applications or patents granted, number of prototype computer programs, etc.

In general, progress seems a misused and misinterpreted term. Its definition, applied to science (and to biology), is a complex task. Alvin Weinberg (1967), a physicist, described three types of criteria for scientific progress; his 'three-level model' includes:

(a) internal impact with its 'own' disciplines;
(b) external (interdisciplinary) impact on other disciplines;
(c) impact on society.

Accordingly various levels of progress can be identified, each with a wider base of impact.

'Internal progress' can possibly be defined in terms of 'counts of output'. Citation indices are useful when comparing two scientists or two institutions within the same discipline, but they are hardly a basis for comparing even two subdisciplines of biology. Still, the recent analyses of Garfield (1982) are extremely interesting. How can we explain, for example, that Sweden's share of the thousand most cited scientists in the world significantly exceeds Sweden's share of the population in the industrialized world, or of the world economy? One conclusion might be that there is strong 'internal progress' within Swedish science.

The second, more broadly based level is 'external progress'. At this level, 'counts of output' seem to be useless as measures of progress. Instead, high ratings (notice and appreciation) by external experts have more weight as indications of progress.

The third level is 'social progress', which can be measured in terms of changes in the values and actions of society. Usually society reacts to its benefit (the introduction of vaccination in the eighteenth century is an example), but sometimes society reacts controversially (for example, the development of biological warfare agents or nuclear weapons); anyway, strong social reaction can be concluded to indicate scientific progress.

The third level of scientific progress defined by social impact involves various elements. One could perhaps be described as a 'paradigmatical element' (Kuhn, 1962), i.e. the elucidation of new scientific facts is included in secondary and primary school curricula. A second element is the introduction of beneficial practices in everyday life (for instance, as far as biology is concerned, improved hygiene, chemotherapy, plant and animal breeding, and food preservation). A third element could be the confrontation of new scientific knowledge with existing moral and ethical values, as well as with external claims to control the 'right to know'. In other words, a challenge to the prevailing norms of society.

'Internal progress' is shown mainly in discussion between scientists within the same discipline. This discussion is documented in specialized scientific journals, esoteric to everyone outside that discipline. 'External progress' penetrates (to use the jargon of publicity) to academic textbooks, and more general journals and periodicals. 'Social progress', however, occurs when new knowledge finds its way into the school textbooks, or is discussed in newspapers and on television and becomes a topic of concern to politicians and the ordinary citizen.

In this paper I intend to analyse progress and its assessment within the framework of the 'three-level model' referred to above, using four examples from microbiology, which is my own field:

(a) procaryotic versus eucaryotic cells;
(b) the procaryotic species concept;
(c) numerical taxonomy; and
(d) plasmids and viruses.

These examples are not necessarily representative, but may reveal different aspects for the assessment of progress in biology. With regard to procaryotic versus eucaryotic cells, there is certainly no doubt about 'internal progress' in terms of new knowledge which has given rise to a fundamental *problématique* and alternative theories of explanation. The essential new knowledge is an understanding of the basic difference in structure and function of two types of cells. The fundamental *problématique* is the origin of those two types of cells, where alternative theories have been presented and supported.

The second example, the procaryotic species concept, can perhaps be presented as an example of failure: the unfruitful and unsuccessful discussion of the bacterial (or procaryotic) species concept that has continued for decades. The existing situation was once described by Stanier and van Niel (1962):

> Any good biologist finds it intellectually distressing to devote his life to the study of a group [of organisms] that cannot be readily and satisfactorily defined in biological terms, and the abiding intellectual scandal of bacteriology has been the absence of a clear concept of a bacterium.

Today one easily gets the impression that the 'intellectual scandal' related to the bacterial species is distressing to the extent that most microbiologists hope for a 'silent burial'.

My third example, numerical taxonomy, describes a case of 'rise and fall'. Few innovations in microbiology have met with such an immediate breakthrough as the introduction of numerical taxonomy (Sneath, 1957). There was an almost exponential growth in the papers published on this topic (increase of 'counts of output' = 'internal progress') in the 1960s.

Twenty-five years later, there is no lack of papers on numerical taxonomy, but prospects for the development of theory, as well as space for new applications in biology, seem to have been exhausted. For a numerical taxonomist, such as myself, this evaluation may look pessimistic. However, among various applications of numerical taxonomy to microbiology, only the field of improved identification techniques is still in a stage of rapid development.

The fourth and final example concerns bacterial genetics, and particularly the impact of continuously increasing knowledge of the extrachromosomal DNA of procaryotic organisms (i.e. DNA included in plasmids and viruses). Research on plasmids and viruses is a dynamic field of biology, and increased knowledge is also a prerequisite for understanding cell evolution and the problems of the procaryotic species concept as well as the principles of numerical taxonomy. In this sense, the example of plasmids and viruses provides a summary of the views presented in the three former examples.

2 THE ORIGIN OF THE EUCARYOTIC CELL

In 1937 a French scholar, E. Chatton, published a book entitled *Titres et travaux scientifiques*. Chatton can be seen as the unknown soldier of modern biology. Apart from that book there are no references to further scientific contributions by Chatton, and even *Titres et travaux* . . . has obviously been studied in the original by only a few scholars (this is indicated by misreference which, in some cases, gives the publication date as 1938). However, Chatton first used the terms procaryotic and eucaryotic organisms, which are now basic concepts in biology. The differences between these cells are now described in every secondary school textbook. For readers whose secondary school education goes back farther than two decades, Table 1 is included to indicate the fundamental differences between procaryotic and eucaryotic cells.

Chatton's new concepts did not give rise to an immediate breakthrough. In the late 1930s there were still enough gaps in our knowledge to leave Chatton's classification of living organisms difficult to understand and to accept. In the 1950s the structure and function of the genome were revealed, as well as the mechanism of protein synthesis. In consequence, a new understanding of the organization of living cells was achieved. With this evidence available, the concepts of Chatton, which in 1937 were merely speculation or prediction, were verified. Although procaryotic and eucaryotic cells are different – R. Y. Stanier, one of the most original and influential microbiologists of this century, has expressed the relation of procaryotes and eucaryotes in the following words: 'The basic diver-

Table 1 *Differences between procaryotes and eucaryotes*

	Procaryotes	Eucaryotes
Number of chromosomes	1	>1
Presence of nucleus	–	+
Presence of specialized organelles (which contain DNA) such as mitochondria or chloroplasts	–	+
Ribosomes	70S	80S
Organellar ribosomes	–	70S
Cell wall contains peptidoglucan	+ or –	–

Mitochondria are the sites for respiration in eucaryotic cells, chloroplasts are the sites for photosynthesis in eucaryotic cells, whereas ribosomes are the sites for protein synthesis. The ribosomes of procaryotes are smaller than those of eucaryotes (S is a unit comparable to weight). Mitochondria and other organelles of eucaryotic cells contain ribosomes but these are identical with procaryotic ribosomes as to their size. The organelles of eucaryotic cells also contain DNA. The organelles, therefore, seem to be rather 'autonomous' units within the eucaryotic cell. Eucaryotes and procaryotes differ also in cell wall structure and composition: eucaryotic cell walls may contain cellulose, chitin and such compounds which are never found in procaryotes.

gence . . . which separates the bacteria and the blue-green algae (the procaryotes) from all other cellular organisms (the eucaryotes), probably represents the greatest single evolutionary discontinuity . . . in the present living world' – yet the fundamental similarity of structure and function of their genomes remains. Accordingly we have to accept an assumption of a kind of relationship or common origin. There is evidence enough to conclude that biological life developed on Earth when conditions were anaerobic (no oxygen was available), and that anaerobic forms of organisms hence represent an earlier stage in biological development than aerobic (i.e. oxygen-dependent) organisms. Anaerobes (also anaerobic photosynthetic organisms) are found only among procaryotes (except for yeasts that are eucaryotes, but are capable of an alternative anaerobic metabolism). Therefore, one may conclude that at least some forms of procaryotes preceded eucaryotes and that eucaryotes hence have had procaryotic ancestry.

As indicated above, the specialized organelles of eucaryotic cells (for example, mitochondria and chloroplasts) are similar in many respects to procaryotic cells. This resemblance has given rise to a theory that eucaryotic cells have developed as a result of endosymbiosis between two kinds of procaryotes. This theory is, however, challenged by another which claims that the organelles of eucaryotic cells have developed due to a differentiation of the original cell. Hence the problem of the origin of the eucaryotic cell has given rise to two competing theories, a theory of 'xenogenic' (externally influenced) development of the eucaryotic cell, and a theory of 'autogenic' (internal) origin of the eucaryotes.

Both these theories face a common dilemma: they are acceptable only upon the presumption of the existence of an original cell, from which the eucaryotic cell has developed either by xenogenic endosymbiosis or by autogenic differentiation. Quite new observations on what has been called *archaebacteria* may indicate that these almost strange forms of procaryotes could constitute the original cell, the 'urkaryote'. It seems too early to express support either for the xenogenous or for the autogenous theory. There are observations in favour of both of them. However, the point is not to decide which is the more suitable theory, but rather to consider that these theories owe their existence to a breakthrough in biological science. In our approach to biological development we are now far beyond the concepts of only a few decades ago. One hundred years after Darwin's death the main problem is no longer 'the origin of species', but the origin of the eucaryotic cell or, even more specifically, the origin of the structures and functions of those cells.

In summary, if the concepts of procaryotic and eucaryotic cells are considered to be progress in biology, how can this progress be identified? Employing the 'three-level model' described above, it seems obvious that procaryote/eucaryote concepts have filtered down to secondary school textbooks. In this respect, the concepts of procaryotes and eucaryotes have introduced a new paradigm, and stimulated new thoughts and views on cellular evolution, which has contributed to an increased knowledge of the very origins of biological life.

This impact is, however, on the level of basic knowledge (or basic research), and one could claim that a social impact or social progress has not been achieved. The various elements of social impact have been touched upon already. Quite obviously, the concepts of pro- and eucaryotes have had a social impact through the 'paradigmatical' element which is strongly manifested in the lower educational curricula.

3 THE PROCARYOTIC SPECIES

In his *Essay on Human Understanding* John Locke (1689) expressed his view of the general species concept: 'The boundaries of the species, whereby men sort them, are made by men.' Half a century later the famous Swede Carl Linné introduced his *System of Nature*, which in his opinion was of Divine origin. Now, in our time, we can of course discuss what Divine origin implies in this context. It could perhaps be taken to mean that the species is an ultimate and indivisible unit in biology, and also a concept that describes an objective entity, independent of man's imagination (not man-made as claimed by Locke). As we know, even in the rational atmosphere of the eighteenth century, Linné's species concept was accepted, whereas Locke's was rejected. This approach certainly had far-reaching consequences. As to microbiology, it can be concluded that, although the 'birth' of microbiology in the 1850s and later was mainly due to the solution of applied 'practical problems' (Pasteur was called on to correct some troubles in wine production and, as a result, described the role of bacteria in lactic acid fermentation in 1857; Koch again, published his classical paper *Uber Wundinfektionskrankheiten* in 1878), the Linnéan approach was, however, also used for the description and classification of procaryotes. Botanical rules of nomenclature were, without any criticism, applied also to microbiology. Louis Pasteur protested against this impact of systematic botany to the new discipline of microbiology when he said that it is more important to know what microbes do, than to know what they are.

In spite of Pasteur's tough declaration, we have since been confronted with the question of what microbes are. It has mainly been discussed in the framework of the Linnéan tradition, the Linnéan paradigm. The present author has made a study of the thoughts on taxonomy by the British microbiologist S. T. Cowan. In my manuscript (1982) I quote Cowan's species concept as applied to micro-organisms:

> A species is a group of organisms defined more or less subjectively by the criteria chosen by the taxonomist to show the advantage and as far as possible put into practice his individual concept of what a species is.

and

> [the species is] a concept, that it is useful cannot be denied, but the user must realize that the species does not exist and is not an entity.

Cowan expresses quite clearly his devotion to Locke's views and his rejection of the Linnéan approach, but still seems to want to retain a 'man-made' species concept for procaryotes. This indicates how closely biology is still attached to hierarchical divisions of nature into classes, groups and basic species units, let them be 'man-made' as stressed by Locke or of Divine origin as believed by Linné.

Obviously this *Ordnung in das Chaos* approach goes further back in the history of science than Lockean or Linnéan thoughts. In fact, it can be followed to primitive structuralism as described by Lévi-Strauss (1962).

So far the species concept is a result of our cultural and scientific tradition, which looks for an exact and static world where even evolution represents a definable force. The alternative is relativism: things do not necessarily exist as such, but in their relation to other things (this philosophy is much in line with modern quantum mechanics). In fact, the species concept, as applied to procaryotes, can be defined as highly relative. Its substance changes due to the relation between the organism under study and the aims and goals of the scientist who studies it. Several different species have been defined. Nomenspecies are fixed to the name used. As a nomenspecies, *Escherichia coli* (the common bacterium of the colon) is defined by all microbial cultures called *Escherichia coli*. Genospecies, again, comprise organisms which are able to exchange genetic material with each other. If applied to *Escherichia coli*, the genospecies would include a number of organisms now classified under various species names. Taxospecies, finally, are groups ('species') produced by some given classification method. The groups thus obtained may include organisms with quite different name designations as well as organisms which do not exchange DNA. Further 'types' of species can easily be described: host species (as defined by specific host) in medicine and plant pathology, serospecies (defined by specific serological features) in immunology, ecospecies (characteristic to specific environments) in environmental science, and technospecies (utilizable for some given technology) in biotechnology, and so on.

Reverting to the problem of progress or failure, it may be concluded that bacteriology (procaryotology) has not resolved its relationship to the species concept. Although, as shown above, the paradigm of procaryote/eucaryote division is generally accepted, the species concept of the 'preprocaryotic era' is still applied to procaryotes. Traditional taxonomy operates with neat dichotomies; it seems, however, that procaryotic taxonomic structures can be better described by the model of 'Russian dolls'.

The classical eucaryote taxonomy implies, first, that taxons on the same hierarchical level of classification are strictly separable from each other, and secondly, that each individual organism ('species' according to classical taxonomy) can belong to only one taxon of the next higher rank (the next higher rank of a 'species' is the 'genus'). The 'Russian dolls' model described above may also be idealized. The reality may show overlap, i.e. a lack of distinct gaps between the taxons, and hierarchical structures may not be operable.

Bacterial (procaryote) taxonomy has existed since the days of Ferdinand Cohn in the 1870s. In spite of more than 100 years of history, bacterial taxonomy has not been very progressive. It has not recognized the very special features of the organisms it considers (the procaryotes) and has therefore been unable to create a novel approach such as an informative taxonomy of the organisms would have developed.

As to the species concept and the classifications based on it, progress in terms of technical development certainly occurs: electronmicroscopy and new tools for chemical analysis have revealed new features of cell-wall structure and composition. DNA base ratios as well as DNA-homology techniques make the direct comparison of procaryotic genomes possible. Chromatography and isotope techniques have added new details regarding biochemical pathways, etc. All of this amounts to considerable knowledge and information which could be utilized in procaryotic taxonomy. However, procaryotic taxonomy as such has not changed. Knowledge and information are organized within the same framework of concepts, principles and general approach as used by the scholars of the last century. This rigid Linnéan framework seems to constitute a serious obstacle in the access to new knowledge and information produced by 'high' laboratory technology.

4 NUMERICAL TAXONOMY

There is one exception to the assessment that bacterial (procaryote) taxonomy has been static and unprogressive. This is numerical taxonomy, which introduced an alternative approach, as well as alternative concepts and principles for bacterial taxonomy. Conventional taxonomy operates with so-called monothetic taxons, which means that one single character may be both essential and sufficient to define group membership. The taxons of numerical taxonomy are, however, polythetic which means that '... organisms are placed together that have the greatest number of shared character states, and no simple character is either essential to group membership or is sufficient to make an organism a member of the group.' (Sneath and Sokal, 1973.) This definition indicates that numerical taxonomy operates by counting the number of shared character states in relation to the total number of defined character states for every possible pair of items in the material to be classified. The number of possible pairs of items increases exponentially as the number of items to be classified increases. For 100 items, for instance, there are $99 \times 100/2 = 4950$ possible pairs, but for 200 items there are already 19 900 pairs. The conclusion is that, to handle large amounts of data, and to sort the figures indicating relative numbers of shared character states, requires the work to be done by computer.

The introduction of numerical taxonomy to microbiology coincided with improved access to computers and software in universities and research institutes. This partly explains the almost immediate break-through in numerical taxonomy. Another contributing factor was the simultaneous development of labour-saving techniques (such as miniaturized diagnostic tests and multipoint inoculators) which allowed performance of a large number of diagnostic tests (i.e. definition of a large number of character states) for many organisms in a reasonable time. Numerical taxonomy can, therefore, be looked upon as a very relevant example of an innovation which becomes possible because a number of conditions were simultaneously fulfilled. The new principles of numerical taxonomy, though theoretically interesting, would have been a daunting challenge from the practical point of view, without advanced computer facilities and improved laboratory methods. This is, of course, an *a posteriori* assessment which also elucidates the general problem of progress assessment: progress cannot be anticipated, although it can, perhaps, be defined in retrospect. Even when one can conclude that many conditions, each of which should stimulate progress, are fulfilled, one can only expect progress, not predict it.

The new possibilities opened up by numerical taxonomy were obvious: materials of manifold size, compared with earlier studies, could be included in taxonomic work. In a short period, numerical methods were applied to a very large number of various groups of bacteria (and even to eucaryotic groups). A further factor which contributed to the increasing popularity of numerical taxonomy was probably the use of binary character states which were familiar to microbiologists.

Numerical taxonomy has produced classifications which agree reasonably well with experience from the application of other techniques. However, as indicated above, the progress has been merely quantitative, involving large materials and a wide range of groups; qualitative progress has been much less obvious. This implies that the role of numerical taxonomy as a technical tool has received much stronger emphasis than its theoretical basis. This has led to a development where numerical classification as a method seems to have exhausted its further possibilities.

On the other hand, the search for efficient determinative and group-separating characters can be carried out by the application of numerical methods which enlarges the scope of numerical taxonomy to cover identification in addition to classification. As shown above, numerical classification is based on the numerical expression of similarity (= the relative number of shared character states). Similarity can be conceptualized as geometrical distance: a short distance implies high similarity, and a long distance low similarity, and hence polythetic groups can be described as aggregates of points – every point indicates an individual organism – in a multidimensional space. Such a model provides a useful framework for identification: an unknown item (organism) would be identified according to where it falls as a new point in a defined space.

The most intensive development in the application of numerical taxonomy in recent years has been in the area of identification. Several interacting reasons can be quoted to explain this trend. First, there has been an increasing need for identification services in public health laboratories, water purification plants, food industries, environmental pollution control, etc. These needs can only be met by automation or semi-automation of laboratory routines, and the introduction of techniques for rapid data interpretation. Secondly, pocket calculators and desk-top microcomputers are fast enough today and can handle even large quantities of material. This has led to the development of 'complete systems' which include ready-made microtest kits and software for the treatment and interpretation of test data.

As to the evidence and arguments for progress or lack of progress influenced by numerical taxonomy, the 'three-level model' is also applicable in this context. At the 'internal level' the impact of numerical taxonomy to general microbial (procaryote) taxonomy is beyond doubt. The number of numerical studies which have been performed is enormous, and the quantity of information produced may exceed the possibilities for concentrated review and analysis. In addition, external progress can be defined, since the breakthroughs in numerical taxonomy have been reflected in many disciplines, even outside biology. Such disciplines include geology, hydrology, and philology. Therefore an impact can be identified which reaches far beyond the boundaries of microbiology, which itself is by no means a narrow discipline.

Finally, numerical taxonomy seems also to influence social progress. The social impact is on the one hand a commercial one, which is connected with the marketing of diagnostic kits and software relevant to the treatment of test results. Besides this, the social impact is revealed by the contribution of numerical taxonomy to the knowledge pool of bioinformatics which is becoming an ever more important element of the approaching information society.

Ultimately, numerical taxonomy can be considered a progressive branch of biology, but it does not constitute a new paradigm, which has changed fundamental thoughts and views. Conversely, it has more or less revealed itself as an innovation which has contributed to conservation of already existing approaches and ideas.

5 PLASMIDS AND VIRUSES

Plasmids are bearers and submitters of extrachromosomal DNA in procaryotes. Hence the role of plasmids is a very central one in genetic engineering attempting to manipulate the DNA of procaryotes. Fred Hoyle (cf. Hoyle and Wickramasinghe, 1981, and Crick, 1983) advocates the view of a continuous transfer of new forms of life from space to the Earth. The basic arguments are interesting but as a microbiologist I have to claim that the relations between 'chromosomal' DNA of procaryotes, viruses and plasmids are not easily explained by this theory. Already the extreme host-specificity of viruses and plasmids seems to exclude ideas of their extracellular origin. A more acceptable explanation is that viruses and plasmids are units which (accidentally) have been separated from the chromosomal DNA. The degree of organization (for instance, ability of reproduction) that viruses (and plasmids) have achieved can be seen merely as an indication of the 'selfish' power of DNA (cf. Doolittle and Sapienza as well as Orgel and Crick, 1980, and the further discussion on this topic in *Nature* 1980; see also Richard Dawkins, *The Selfish Gene*, 1982) and of its flexibility in finding the means to survive and to reproduce. This kind of discussion brings one back to the problem of cellular evolution. Besides the problem of the origin of the eucaryotic cell, which has already been discussed above, we are also confronted with the even more fundamental problem of the origin of the procaryotic cell (or the origin of DNA itself). Viruses and plasmids could be considered as indications of the capacity of DNA continuously to give rise to subunits of itself which are organized to survive independently.

As to the plasmids, we know that in the symbiosis between leguminous plants and bacteria, the nitrogen-fixing ability is bound to bacterial plasmids. We also know that this important state of symbiosis obviously represents a rather recent state of evolution. Both leguminous plants and their bacterial symbionts are able to survive in a non-symbiotic state, providing that mineral nitrogen is available in the soil. Possibly the readiness for symbiosis (in the absence of mineral nitrogen) represents, from the side of the bacteria, an evolution of pathogenicity. Nitrogen fixation, which results in the production of ammonia, was perhaps originally a means to defend an ecological niche. Certain plants succeeded in adapting themselves to this circumstance (the high concentrations of ammonia). Contrary to what is generally concluded, the adaptable counterpart in legume/rhizobium symbiosis is the plant, not the bacterium.

Apart from this conclusion, the nature of the nif-gene, which determines nitrogen fixation, needs elucidation. This approach implies that a completely new taxonomy – a 'plasmid taxonomy' – may be introduced. Such a taxonomy would perhaps indicate relationships between nitrogen-fixing clostridia (which are anaerobes, and which thus constitute a rather ancient form of procaryotes) and the 'modern' legume-symbiotic rhizobia. The problems of present-day biology are no longer those of species. Modern biology explores the evolution of cells, of their components, of their organization and of their function.

We can ask ourselves if this development is progress or not. Certainly, it is progress in the sense that new paradigms (like Hoyle's suggestion of the extraterrestrial origin of viruses) are continuously suggested, discussed and tested. The new comprehension of plasmids and viruses fulfils the criteria of the third level in the 'three-level model', i.e. that knowledge has made inroads into secondary school textbooks and can be used in a socially beneficial way. However, the use of plasmids in genetic engineering has brought into focus moral and ethical questions. It should also be considered that much of existing knowledge is kept secret due to reasons of national security or commercial interests. These last considerations bring us into conflict with moral and ethical norms, and raise the question as to whether claims of secrecy can be seen as criteria of scientific progress.

6 SOCIETY AND SCIENCE

Assessment of progress deals with progress. Criteria for progress are defined from an analysis of success. This implies that we try to derive a basis for prediction of progress from what we assess as a useful or successful future output.

This conclusion may seem controversial, but scientific progress is, at least to some extent, a matter of convention or agreement. Scientists and scholars agree to accept some theories, descriptions or conclusions as progress. Progress is then what is considered as progress in the prevailing and mainly accepted frame of reference for a given scientific discipline. On the other hand, there obviously occur situations where *a priori* conventions have been invalidated by some *a posteriori* conclusions. Analysis of such cases usually shows that a significant prerequisite for progress has been a combination of existing knowledge from various sources and disciplines.

Looking at the cases presented above, the procaryote/eucaryote paradigm constitutes a good example of combined knowledge resulting in progress. The same certainly concerns the plasmid/viruses case. The case of numerical taxonomy is also an example of combined knowledge which constitutes both theoretical and practical innovations. However, innovation is not a synonym for progress. As concluded above, the progressive element of numerical taxonomy is disputable, and it seems to be an example of 'unprogressive innovation', which is very interesting in itself, and perhaps worth more extensive study. Finally, the species concept is a case where no efficient 'combination of knowledge' has taken place, and where no particular progress is identifiable.

A rather unorthodox criterion for scientific progress has been referred to above: lasting (accepted, agreed) scientific progress percolates down from scientific reports to academic textbooks and thence to those of secondary or primary schools. Application of this 'contribution to education' criterion shows that the procaryote/eucaryote paradigm, as well as knowledge of plasmids and viruses, has found its way to the secondary school level. Numerical taxonomy is well established at the level of academic textbooks, whereas the discussion of the species concept is still limited to so-called scientific reports and reviews.

Attempts to compare even distant scientific disciplines have created seemingly common categories for comparison such as cumulation, integration and innovation. In biology (or microbiology), cumulation is relevant, knowledge is additive, yet the problem remains unsolved. The species concept is an example of cumulative knowledge and problem stability, quantitative information does not solve an existing (eternal) problem. The integration concept, in the sense of combination or integration of knowledge, has been discussed above. Obviously the procaryote/eucaryote paradigm is, among the cases presented, a good example of successful integration. Numerical taxonomy, again, can well be chosen as an example of innovation. However, as has already been pointed out, innovation does not necessarily imply progress. So far, scientific progress cannot be clearly defined in terms of cumulation, integration and innovation. A paradox of rational science is the irrationality of its progress, which is possibly due to social expectations which go beyond purely scientific frames of reference but still exert a strong impact on science.

Scientific progress, also in biology, has been defined above in the somewhat narrow context of acceptability in general education. A wider context is the broad impact on society. Among the cases presented above, only plasmids and viruses which find an application in genetic engineering seem to represent a broad impact. The last-mentioned case is also a useful example of a reverse impact, i.e. that of society on science. The directing of science by social expectations and claims includes significantly beneficial effects, but also obstacles to scientific progress.

A novel by the Swedish author P. C. Jersild (*En levande själ*, 1980) elaborated the fiction of using human brain tissue to speed up and to maximize the capacity of computers. Today 'biochips' are no longer a matter of science fiction but known as possible 'high technology' of the next decade. However, the active material in the 'biochips' is by no means human brain tissue, but molecules of bacterial origin (which was in fact anticipated by Jersild). Jersild's outline of the problem, i.e. that human society expects and wants inhuman science, went extremely deep into fundamental moral and ethical questions connected with the ultimate limits of human utilization of biological life.

In the context of a discussion of scientific progress it seems possible to formulate the following questions: should it be considered scientific progress when scientific thinking and the application of scientific results are in conflict with the ethical values and moral codex of our society, or should we regard ethics and morals as obstacles to scientific progress which must be overcome? These questions are not new ones; in generalized terms, they have been discussed by classical authors like Dostoiewski. Dostoiewski (in *Crime and Punishment*) gave Raskolnikov at least some hope, but what is left for science if it destroys the fundamental norms and values of our culture. Jersild has recently published another novel, *Efter floden*, where he tries to describe what our culture would be like after a nuclear war. It is a depressing story which must impress upon every reader that there are limits which simply cannot be exceeded. The specific problems of morals and ethics in (micro)biology have also been discussed. It has been emphasized that we are at a crossroad: if our purpose is to create a new order of organisms engineered by man, we must also be prepared to face all the repercussions which will be more far-reaching than we can imagine, and which cover all aspects of human society.

Criticism can certainly be raised against the above discussion. It is true, of course, that an assessment of science and scientific progress in terms of morals, ethics, religion, etc., is non-scientific. Scientific progress should be assessed in scientific terms and by scientific methods; morals and ethics are different things which should be assessed according to moral and ethical considerations. This *problématique* is 'eternal' and probably unclassifiable because the very concept of science includes moralistic elements (just to mention the requirement of 'truth', a concept which is perhaps more easily defined in moral than in scientific terms). These moralistic elements are also encountered in any discussion of secrecy and the outside control of science. The problem of intellectual property or 'who owns scientific information' is treated in a consistent and constructive way in a recent essay by Nelkin (1982).

Claims for secrecy are mainly two, namely national security or national economy (maintenance of commercial competition). In both cases the background reasons can, of course, be interpreted as acceptable. National security is considered as of the utmost priority in every political system, and it is also considered as fully legal to protect taxpayers' or shareholders' money. Perhaps classification as secret can be taken as verification of the impact of scientific effort (there is no need to claim secrecy for, or claim control over, non-existent knowledge). Therefore, if an item of knowledge is kept secret, it could be taken as an indication that it is both useful and acceptable.

The above considerations give rise to new questions:

Is classification as secret an indication and a verification of scientific progress?
Will secrecy, in the long run, be an obstacle to further scientific progress or does it foster a scientific élite which stimulates scientific progress?
Does science flourish better in the atmosphere of the strong goal-orientation of military or commercially useful research, or is non-oriented 'free' research more stimulating?

In discussing the answers to these and other similar questions, the first obvious conclusion is that classification as secret – for either security or commercial reasons – is a non-scientific issue. It has already been pointed out that science involves moralistic elements; since national security and commercial competitiveness can be considered as morally 'good' or acceptable as social goals, science can and perhaps should be used to achieve national security and wealth. On the other hand, there seems to be no scientific approach to the selection of those research projects which are thought to enhance 'good' and 'acceptable' social goals.

Secret research is usually considered obscure by scientists who are not themselves involved in secret projects. Those who are involved in secret, classified research are not, by definition, allowed to speak on the matter. The classified scientists probably find themselves well motivated, well paid, and do not have to worry about expenditure even when it exceeds budgetary limits. Things will be regarded in order as long as the research is considered useful. Accordingly, the classified scientist (involved in either military or commercial research) has to make himself continuously useful and necessary. His unclassified counterpart or colleague has to express himself in other terms and finds himself in the 'publish or perish' race. Actually, the incentives are very much the same: a scientific career, continuous financial support, personal recognition. This implies that the individual scientist utilizes research to a certain degree to promote his own personal satisfaction in life and his financial income. This raises a further question in the context of social impact on science: can the personal success of individual scientists be regarded as an indicator of progress of the successful scientists' disciplines?

The above survey has not dealt very comprehensively with biology; the points emphasized and the questions raised are general ones. However, most of the questions dealt with are especially relevant in progressive fields of science (such as genetic engineering and biotechnology). At the present stage of development the expectations exerted by society are rather demanding. Therefore, the whole spectrum of problems formulated above – the confrontation of scientific progress with moral norms and social values, secrecy as stimulus or obstacle, scientific activity as a means to support individual satisfaction and economic success, etc. – is also most relevant to biology. Possibly problems of this kind are, at least at present, more accentuated in biology than in various other disciplines. These problems are, on the other hand, quite general and some of them, like that of the individual scientist as a career-hunter, have already been described by Max Weber.

Society can perhaps escape science, but science can never escape society. Society puts the pressure of its expectations and claims on science. This introduces, as has been indicated in this essay, several new aspects of scientific progress. However, although society perhaps can escape science, there is also a recognizable impact of science on society. As far as (micro)biology is concerned, it is difficult to translate impact into clear terms of scientific progress. One conclusion, which lies close at hand, is the scientific impact in developing countries. Again, how can we assess the success of, say, vaccination campaigns, introduction of biological insect control or legume seed inoculation to increase nitrogen fixation? Is it already an indication of progress that measures like these are undertaken or should there also occur relevant evidence for the success of these measures? This is certainly not a rhetorical question: it is a most relevant question to many United Nations agencies, as well as to national aid agencies. It has been reported that the Head of the Kenyan State, President Daniel Arap Moi, has decided to apply inoculation of legume seeds on his private farm. Does this represent scientific progress, and who is more powerful in influencing the public opinion, and thus increasing application of legume seed inoculation, President Moi or his scientific advisers?

7 CONCLUDING REMARKS

The present author cannot claim to have presented views that are generally accepted within his discipline, which is microbiology. The views and conclusions are highly personal, and possibly provocative, and aim to stimulate discussion.

The main contribution, in the author's opinion, is the 'three-level model' and its corollaries. This model comprises three levels: internal progress, external progress and social progress. As to internal progress, the author suggests that, at this level, progress is what influential scientists decide it should be or what is mutually agreed to be progress. Citation indices and career histories reveal nothing more than prevailing devotion to, or predomination of, certain 'schools' or institutes.

The present essay does not touch very much on external progress. External progress implies general acceptance from the scientific community. When philosophers become interested in sociobiology or when social scientists become aware of the consequences of genetic engineering, an external impact or even progress is obvious. It seems that cross-influences of this kind have been less intensively studied, although one of the fundamental indications of scientific progress, the cross-fertilization or integration of knowledge from different sources, could be regarded as external progress.

Finally, social progress has perhaps been dealt with in the most detail in this essay. Four (or five) dimensions of socially related scientific progress have been described (impact on education, 'beneficial' social utilization, confrontation with moral and ethical norms, claims for secrecy, and even the success of individual scientists). Except for the dimension of social benefit, they will probably give rise to controversial and critical opinions. The present author feels, however, that the dimensions of social progress of science (biology as well as other fields of science) represent the most fundamental problems facing science in our present society.

8 EXPLANATORY COMMENTS

Some concepts included in the above text may require further explanation for those readers who do not have an educational background in biology. Brief explanatory notes are therefore added.

The genetic information of a procaryotic cell is carried on the structure termed the procaryotic chromosome. It consists of a double helical DNA (deoxyribonucleic acid) molecule which in some procaryotes is circular. This molecule is referred to as chromosomal DNA of the procaryotes. This chromosomal DNA of the procaryotes has its counterpart in the DNA which is included in the nuclear chromosomes of eucaryotes, but the structures are essentially different. The chromosomal DNA of procaryotes resembles in structure the DNA present in the organelles (for example, mitochondria and chloroplasts) of eucaryotes. Many procaryotes can also harbour small circular DNA molecules which are capable of replication, independent of the chromosomal DNA. These entities of 'extra-chromosomal DNA' in procaryotes are referred to as plasmids. Several interesting and important characters of procaryotes, such as resistance to antibiotics and the ability to fix atmospheric nitrogen, are coded for by plasmids. Since procaryote cells can exchange plasmids, they may – even under the conditions prevailing in nature – transfer given characters or features from one organism or one population to another. In genetic engineering plasmids can be utilized as vectors which transfer genes (and thus characters and features) from one organism to another.

Compared with plasmids, viruses represent a further stage of organization. They consist of DNA or RNA (ribonucleic acid) which is 'protected' by a protein coat. When virus particles (for example, the virus causing our common cold) invade susceptible cells they lose their protein coat and start to utilize the host cells' genetic machinery for replication of their own DNA (or RNA). Since viral DNA, like plasmid DNA, represents an extrachromosomal DNA element in the host cells, it can be utilized as a vector for artificial gene transfers.

The term symbiosis implies that two organisms have developed an interdependence which is beneficial for both partners. In the case where one partner of the symbiosis lives within the cells of its host this relationship is called endosymbiosis. There obviously exists a kind of symbiosis between man and some of his intestinal bacteria. These bacteria do not, however, invade our tissues or cells but are attached to the intestinal walls. This symbiosis can therefore be described as ectosymbiosis. The xenogenic theory of eucaryotic origin suggests that the organelles of eucaryotic cells have their origin in procaryotic endosymbionts.

The DNA molecule is a double helix. The two strands are connected by 'bridges' of two bases. There occur only four bases: adenine, thymine, guanine and cytosine. In the 'bridges' adenine always combines with thymine, and guanine with cytosine. Accordingly, the amount of guanine + cytosine in relation to the total amount of bases constitutes a rough character of the DNA in question. If the guanine:cytosine ratio (or GC ratio) of two organisms is considerably different, one can conclude that there is no close relationship between those organisms. On the other hand, if the GC ratios are closely similar, this does not necessarily indicate taxonomic similarity (for example, man and some bacteria have rather similar GC ratios).

The bases mentioned in the preceding paragraph, and the order in which they occur in the DNA strands, define the genetic information. It is therefore obvious that closely related organisms possess highly homologous DNA. It is possible by some techniques to determine the homology of the DNA from two different organisms. Comparison of DNA from two human individuals will give a homology close to 100 per cent, but a comparison of the DNA from two bacteria which are not taxonomically related will drop the homology to almost zero. DNA homology determination, although still rather laborious, is therefore a useful tool in procaryote taxonomy.

The concepts of hardware and software have been introduced from computer terminology. Hardware represents the computer itself and further necessary machinery, whereas software consists of the programs and other immaterial input to get the computers working. These terms have since been adopted to other comparable fields. In connection with the identification of micro-organisms, hardware consists of the tools and utensils required (such as the kits to perform necessary tests) whereas software treats the primary data (catalogues, microcomputer programs, etc.)

Comment

R. B. CLARK

1 INTRODUCTION

Professor Gyllenberg examined the impact of scientific progress at three levels: internal, meaning that the innovation or discovery has influence within its own discipline; external, meaning that it has an impact on other scientific disciplines as well as its own; and third, one that has an impact on society at large. There are categories of social impact: the scientific innovation is absorbed into the general culture, at least to the extent of being included in school textbooks; it introduces beneficial new practices into everyday life; or, most significantly, it confronts existing moral or ethical standards.

Professor Gyllenberg then tested these criteria against four important topics in his own discipline, microbiology. Of these, one had made little progress; a second, numerical taxonomy, has had a marked influence within its own discipline, but has proved to be an essentially conservative development. The two remaining questions, on the other hand, have been the basis of great progress and even some social impact to the extent of being incorporated into school textbooks.

I recognize the value of Professor Gyllenberg's analysis, for, as he intended, it is certainly provocative. It is by no means evident that the criteria he advocates provide a reliable diagnostic tool for identifying or measuring scientific progress. Nor does the other criterion he suggested – that scientific progress is what senior scientists at the time say it is – seem any more reliable, though perhaps that is a more accurate picture of how we recognize progress at present.

My doubts are best illustrated by citing some examples, though I will approach the subject from the opposite direction to that of Professor Gyllenberg. What has been the scientific importance of biological innovations that clearly have had a social impact? What biological discoveries have been misjudged by contemporary opinion about their potential to exert an external influence? How successful has been the identification of innovations within a discipline?

2 INNOVATIONS WITH SOCIAL IMPACT

Foremost among biological innovations that have had a major social and moral impact must be the theory of evolution by natural selection. It rapidly became a matter of intense public debate which, surprisingly, is still not over. It had an impact on the moral climate of society; it provided a concept which has been adopted, consciously or not, in most aspects of human thought and activity; and it provided a new conceptual framework for all zoological studies. It meets all Professor Gyllenberg's criteria of progress in a striking way, but few scientific innovations have such universality, and more recent examples that I shall quote appear to be of a different order of magnitude. There can be no doubt, though, about the impact on society at large and its moral climate of such innovations as the discovery of antibiotics and their introduction into medical therapeutics; the development and widespread use of organochlorine pesticides, specifically DDT; the development of a reliable oral contraceptive; and recent developments in genetic engineering.

Antibiotics rank with antiseptics and anaesthetics as boons to mankind that have made the second half of the twentieth century infinitely more comfortable than any previous period in human history. They are a major contribution to the ability of modern medical science to prolong life. But this has led to a new consideration of the quality of life and raised the question of at what point should artificial life support systems be withdrawn in terminal illness, and who should decide when they should be withdrawn. In the United Kingdom, at least, this debate in the late 1970s led to a redefinition, in law, of clinical death. That seems a profound social impact for any biological innovation.

Pesticides, similarly, have been a boon to mankind. Without DDT there would have been major epidemics of typhus and other insect-borne plagues in Europe during and immediately following the Second World War. The World Health Organization based a highly successful malaria eradication programme on DDT. But in 1962 Rachel Carson published *Silent Spring*, which, if it does not mark the start of the environmentalist movement, was instrumental in giving it a wide and popular following. DDT does not kill only malaria-transmitting mosquitoes but also bees and butterflies. It is persistent and accumulates in food chains and was responsible for the reproductive failure and population decline of birds of prey in Europe and North America and of pelicans on the California coast. DDT is not known to harm humans but there were legitimate concerns about the long-term effect on the natural environment of the use of these persistent pesticides and, by extension, of all forms of intensive agriculture. By now the environmentalist movement has become a serious political force, most conspicuously in Germany, though not confined to that country.

I leave it to experts to decide if the ready availability of reliable contraceptives has been responsible for a revolution in moral attitudes towards sexual behaviour in western society. This has certainly been claimed, though there is also the view that it is not so much sexual behaviour that has changed as the freedom with which it is discussed.

Nevertheless, the introduction of oral contraceptives has posed a serious moral problem in a number of western countries, among Roman Catholics who find themselves at variance with the teaching of their Church.

It is perhaps too early to judge the full social impact of genetic engineering. The economic impact is already large and the moral impact may yet be to come. It is one of the very few examples I know of the practitioners of a new science being sufficiently concerned about its possible consequences to call a moratorium, even if it was temporary and unsuccessful, on further work on the subject. Someone had looked inside the crack in the lid of Pandora's box and tried to close it again, but too late.

There can be no doubt that these biological innovations have had a profound social impact and posed moral questions: of life and death, leading to a revision of the legal definition of death; of man's relationship with his environment, leading to political activity which can shake governments; of sexual relationships, leading to conflict with Church authority; of a crisis of confidence in our ability to control a Frankenstein which we have perhaps created. I doubt if even the theory of evolution by natural selection had so great an impact on society.

But the scientific importance of these innovations to their own disciplines was very varied. The discovery of the antibiotic properties of penicillin appears to have been a major advance, but DDT was known as a chemical in the last century and its introduction as an insecticide around 1940 was only the result of patient and routine screening of likely compounds in the search for new pesticides. Understanding the reproductive endocrinology of female mammals was an essential precursor to the development of oral contraceptives. Gaining that understanding was a major achievement, but one achieved by steady progress over several decades, accompanied by a very large volume of research into steroid chemistry. It would be difficult to identify a particular point in this investigation where a crucial discovery was made. Equally, I suspect it is a matter of personal judgement as to when the critical step was made in the development of genetic engineering. The development of these techniques, however, has provided a powerful research tool for pure science in a way that the development of DDT and oral contraceptives has not.

How a new development in science is evaluated depends very much on one's viewpoint. The social impact of a scientific innovation does not have a very close relationship to the consequences of the innovation for the development of the science itself. It may be noted, too, that social assessment is fickle: DDT was regarded as an unqualified benefit to mankind in the 1950s, but in the 1960s was regarded quite differently. The changed evaluation may follow upon considerations of social science or economics, but has no implication for chemistry, biology or the other basic sciences.

3 CROSSING DISCIPLINARY BOUNDARIES: MISJUDGEMENT OF PEER REVIEW

It would probably not be difficult to catalogue misjudgements by the scientific community of the value of new developments, particularly in assessing how widespread their influence is likely to be. I select two examples for the lessons they illustrate. The first is numerical taxonomy, cited by Professor Gyllenberg, which had a much narrower impact than was anticipated; the second is neurosecretion, which, far from being accepted as an advance when the phenomenon was discovered, attracted disbelief and hostility for a quarter of a century. Yet it is now clear that it was a discovery of first importance in a variety of biological disciplines, initiated many new lines of enquiry, and may yet achieve the distinction of having a social impact.

Taxonomy is concerned with classifying organisms in a rational way. Since classification is essential in many fields of human activity, there is no reason, in principle, why new insights into biological classification should not have widespread application; at least they should find ready acceptance in the biological sciences. Numerical taxonomy appeared to offer such a new advance. It avoided subjective judgement, which had always been an unsatisfactory feature of biological classification, and it had biological validity because it depended on quantifying the genetic differences between related organisms. Yet, while numerical taxonomy has been eminently successful in microbiology, the field in which it was first introduced, it does not appear to have had much influence in other branches of biology, nor does it seem to have had any social impact. Here I think I part company with Professor Gyllenberg. It is true that we have all become more numerate in our classification, but that does not seem to me to be attributable to the introduction of the techniques or concepts of numerical taxonomy. At any rate, numerical taxonomy failed to make a permanent mark in my discipline, which is zoology.

I believe there were two reasons for this failure. First, bacteria are simple organisms and a reasonably complete enumeration of the characteristics of each is a manageable task. Higher organisms are infinitely more complicated, and whatever the theoretical merits of numerical taxonomy, it is impracticable for botanists and zoologists to apply it to their own work, except in very limited circumstances. But more important was the depressed state of taxonomy, particularly zoological taxonomy, which militated against the acceptance and development of new ideas. The great success of molecular and cell biology in the 1940s and 1950s naturally attracted clever, young scientists entering the profession, as well as substantial financial support. Success breeds success, and taxonomy, then an unfashionable subject, suffered from a lack of good recruits and lack of finance. It was not fruitful ground in which to sow new seeds.

A completely contrary fate met the discovery of neurosecretion. In 1928 Ernst Scharrer described nerve cells in the brain that secrete hormones into the blood stream. They form part of the pituitary gland and are the bridge between information received by the brain through the senses and the body's response to environmental changes. It was subsequently shown, in large part by Scharrer and his wife, that neurosecretory cells occur in all animals that have a nervous system and in lower animals they constitute the entire endocrine system and control such fundamental processes as growth and reproduction. The discovery of these cells has led to the study of the role of peptides in the nervous system and this promises to lead to a better understanding and perhaps therapy of diseases such as schizophrenia. It is hard to exaggerate the importance of Scharrer's discovery, but far from it being recognized for the advance that it was, it attracted disbelief and even hostility from the scientific and medical establishment for almost 30 years.

The reason for this failure to recognize the significance of an important discovery was not lack of interest in the subject or lack of distinguished scientists working in the field. Endocrinology, and particularly the structure and function of the pituitary, was a buoyant subject, vigorously pursued in many laboratories in Europe and North America by both medical and pure scientists throughout this period. The subject showed many and rapid advances, and attracted good new scientists and strong financial support. Unfortunately, so far as the pituitary was concerned, collective scientific wisdom was focussed on one particular approach, and Scharrer's discovery, far from being an advance, was regarded as an irrelevant distraction from the mainstream of research. In consequence it was ignored.

4 IDENTIFYING PROGRESS WITHIN A DISCIPLINE

Whatever influence a new discovery may have on science at large or on society, its immediate impact will be within the discipline in which it was first made. It is not very difficult with hindsight to identify what have proved to be important advances in the development of a branch of science. But this enquiry is interested in finding hallmarks of progress that have predictive value rather than in engaging in a retrospective survey of scientific activity. Research councils naturally wish to use their research funds wisely and it would greatly assist them if they knew which horse was going to win a race before placing their bets. Experience of past successes and failures at picking winners – for grant-awarding committees as for punters at the race course – does not encourage the view that either have found an infallible system.

To judge from previous experience, for a scientific innovation to flourish and develop, a number of conditions must be satisfied.

(a) The innovation must have the qualities claimed of it; in other words, it must be right. Numerical taxonomy was a successful innovation in one discipline, but did not have the potential for widespread application that its more enthusiastic proponents claimed, and it proved to be impracticable as a tool for the classification of organisms with the complexity of higher animals.

(b) The time must be right for the innovation. Darwin's *Origin of Species* was an instant success. There was nothing new in the idea of evolution. It had been envisaged by the Ionian school in classical Greece and Charles Darwin's own grandfather, Erasmus Darwin, had written a lengthy treatise, in verse, on the subject. All that was lacking was a convincing mechanism by which it happened, and natural selection provided it. It was seized upon immediately. Mendel's discoveries relating to genetic inheritance received no such recognition, through lack of an appropriate audience, and the initiation of the modern science of genetics had to await the rediscovery of Mendel's work by Morgan 40 years later. Scharrer's discovery of neurosecretion did not lack an audience, but had to wait 30 years for acceptance until the current mainstream of research had proved mistaken.

(c) If the time is right for a discovery, it will certainly be made by somebody. The first statement of the theory of natural selection was, in fact, a joint communication by Charles Darwin and Alfred Russell Wallace, who had arrived at the same conclusion quite independently of each other. If Crick and Watson had not produced their model of the structure of DNA, Wilkins in London or Pauling in Pasadena would surely have done so. In rapidly moving fields of research, it is commonplace for the fear of being 'scooped' by a rival laboratory to be a serious concern of researchers.

These conditions apply in fashionable subjects where there is wide knowledge of the questions that need to be answered and an informed audience that will recognize a satisfactory answer when it is presented. That is, the discipline is in a progressive phase: it has a substantial number of practitioners and, most importantly, it is exciting and attractive to bright, young scientists. In these circumstances there is no difficulty in identifying where research effort is needed, or in recognizing progress where it is made, and usually there is no doubt which scientists or laboratories are likely to achieve that progress. Research councils know where to place their money and have no difficulty in picking winners. But fashionable subjects suffer from two deficiencies: they become popular, and they tend to outstay their welcome.

When it becomes clear that a particular area of research has made a breakthrough, it attracts older researchers in allied fields who have no firmly fixed research of their own and it attracts many young recruits to what appears to be the forefront of science. The subject becomes a bandwaggon. Many of the recruits, young or older, are not in the first rank of researchers, and the subject, after its peak has passed, becomes burdened with a substantial fringe of frankly second-class investigators. This is true of some areas of cell biology and is conspicuously true of much ecological and pollution research.

A subject that has become fashionable in this way also inevitably develops into a long-standing commitment. It has attracted a large number of scientists, distinguished and not so distinguished, who will expect support and laboratories, equipment and staff that cannot easily be diverted to other subjects. The bandwaggon has acquired a momentum of its own and there is a strong likelihood that it will continue to receive a high level of support after it has ceased to justify it.

In contrast, research in unfashionable fields or research in fashionable subjects which runs counter to the direction favoured by collective wisdom usually has the greatest difficulty in attracting finance. I am not sure how many unknown geniuses are left to wither from lack of support – ugly ducklings have a habit of developing into ugly ducks – but the system of research council grants, with peer review of research proposals, is not well placed to discover them.

A second difficulty is in the recognition of innovative, young scientists at an early stage in their career. There is some truth in the claim that scientists have all their really original ideas before the age of 30 and spend the rest of their lives, and much of the time of their students, validating them. At present, young, innovative scientists are largely dependent on finding a patron, in the form of a senior scientist of some standing, to nourish them until they have become accepted. This dependence has the danger of parasitism of the younger by the older, but more importantly of a censorship of new ideas. There are fortunately many notable exceptions and most senior scientists are concerned to support and encourage the original thought of younger colleagues, but the dangers remain.

The problems of identifying unconventional, but potentially fruitful new developments in a science are not serious when the economic climate is healthy and research support is readily available. Obvious lines of research can be generously supported and there is surplus money to support, perhaps rather uncritically, a variety of unconventional research in the hope that some dark horses will prove to be winners. Most will fail, but some will provide the basis for perhaps the most significant new advances in the science. In times of financial constraint, however, it is the unconventional research and the young scientists with original ideas that are most vulnerable. Until recently, in Britain and some other countries, the universities were able to provide for the essential substratum of research, by making research, as well as teaching, part of the terms of employment of university staff, and by providing time and a limited amount of financial and technical support for the research of its members. This gave independence to the young scientists and an opportunity for them to go some way to testing and validating their ideas before having to submit them to review by more senior and established scientists ('peer' review) when applying for greater support from research councils. A considerable fraction of university research, much of it of great future importance, was, in fact, conducted without recourse to research councils. By now, little experimental research is possible in universities without the help of research grants from external sources.

One is not encouraged to believe that a centralized system, represented by the research councils and based as it is on peer review, is very reliable at identifying anything but the obvious for support. It will not have escaped attention that, fortuitously, more than half the major advances in biology that I have discussed were achieved without the benefit of peer review. I will not hazard a guess as to the number that would have been successful under such a review system.

Professor Gyllenberg raised the issue of the confidentiality that exists in some areas of commercial and defence research. Both attract high levels of support which, in aggregate, dwarf the finance devoted to the non-commercial, civilian research we are discussing. Both commercial and defence research are extremely wasteful. Commercial research invests large sums in routine investigation, most of it with little return, as, for example, in the screening of likely compounds for their insecticidal or pharmaceutical properties. Military research, although I speak with no inside knowledge beyond that afforded by the newspapers, is even more wasteful. The secrecy attached to these activities is not related to their scientific importance, though that is not to say they have none, but to take Professor Gyllenberg's point, they are regarded as socially important activities; at least, our societies devote considerable resources to them. In the absence of any sure way of identifying productive new ways to proceed, the wastefulness of providing for exploration, much of which may prove fruitless, is essential in these investigations as in any others. The universities used to provide a testing ground for fruitless and fruitful investigations in non-commercial, non-military science, but that has been largely eliminated. Research councils, operating on a totally different financial scale from commerce or defence research, are now seeking to eliminate waste, to avoid going down blind alleys, of foreseeing the future prospects of all new ideas. It cannot be done.

Assessment of progress in medicine

H. R. WULFF

EACH week, thousands of scientific papers are published in a multitude of medical journals, and we characterize those which we happen to read as good or bad, important or unimportant. What are our yardsticks when we make these assessments, and what are the yardsticks of society when it assesses progress in medical science? Which factors inhibit medical progress, and are we able to remove some of the obstacles? Such questions are extremely complex because of the diversity of medical research, and in order to avoid an abstract discussion, I shall approach this topic by means of a concrete example.

I AN EXAMPLE: THE DEVELOPMENT OF MEDICAL TREATMENT OF PEPTIC ULCER DISEASE

The treatment of peptic ulcer disease has been revolutionized during the last seven or eight years and I shall attempt to analyse the background of this development in some detail without being too technical.

The story begins at the start of the nineteenth century when doctors in France for the first time identified diseases with anatomical lesions. This is a commonplace idea in medicine today and it is difficult for us to imagine how revolutionary it then was. Gastric ulcer (ulceration in the stomach) and duodenal ulcer (ulceration in the duodenum) were established as disease entities, and later that century the first comprehensive clinical descriptions were published. According to a Danish report, for instance, gastric ulcer disease was extremely common around 1880, especially among young girls from the working classes, and many of the patients died from complications like bleeding or perforation. In contrast, duodenal ulcer disease seems to have been very rare at that time, but the picture was soon going to change, because after the turn of the century gastric ulcer disease in young women almost vanished, and duodenal ulcer disease became very common among young men. Today, the picture has changed again, as duodenal ulcers are found mostly in middle-aged men, whereas gastric ulcers are most often seen in elderly people, both men and women. Obviously, these changes can only be explained by altered living conditions, and a few epidemiologists have tried in vain to determine the causal factors. In this area there has been little progress, and we are still unable to prevent ulcer disease.

Progress was only made when doctors in the later part of the nineteenth century began to study the physiology of the digestive tract, and it was soon discovered that ulcer formation is the result of autodigestion. Hydrochloric acid and pepsin are necessary constituents of the gastric juice for digesting proteins and, if the normal defences are insufficient, acid and pepsin may cause ulceration in the digestive tract itself. Since this discovery, the neutralization of acid or the inhibition of acid production have been the mainstay of medical treatment, but until recently the available remedies were very ineffective.

The breakthrough happened at a completely different frontier of research. In the 1920s it was found that fluid from damaged tissue contained a substance (H-substance) which was later found to be histamine, and this discovery led to an abundance of research. It was found that histamine plays an important role in allergic reactions, and it was shown that the injection of this substance has a variety of effects, including the stimulation of acid secretion in the stomach. In the 1940s a number of antihistamines were produced for the treatment of allergic conditions and it was found that these drugs neutralized all the effects of histamine except the effect on acid secretion. This was the observation which led to the development of modern ulcer treatment. If it had proved possible to produce antihistamines which blocked the effect of histamine everywhere else in the body, it might also prove possible to synthesize other antihistamines which only blocked the effect on the stomach. On the assumption that histamine plays a physiological role in acid production, such drugs might prove useful for treating ulcer disease. This bold idea (which is based on the theory of specific receptors and substrate competition) proved successful. Most of the research was done at a particular pharmaceutical company and in the 1970s the first antihistamine with a specific effect on histamine-induced acid production was ready for testing. So-called double-blind clinical trials proved that the drug greatly enhanced ulcer healing, and it is now used in the clinical routine.

Concomitantly, surgical techniques have improved, and it is also possible to cure most patients with duodenal ulcers by simply severing those nerve fibres which induce the stomach to secrete acid.

2 THE IDEA OF PROGRESS

The very idea of progress or advancement presupposes a goal-directed activity, and it is not possible to discuss progress in medicine unless we consider the aim of medical research.

Medical ulcer research proved a success in so far as it resulted in effective treatment of ulcer disease, and in general terms we may state that *it is the aim of all medical activity to promote health and to eliminate illness by prevention or cure*. Medical activity always has a practical purpose and from that point of view medicine has more in common with engineering than with physics.

In contrast, biology is a basic science, as it is the ultimate aim of biological research to acquire true knowledge about living organisms, regardless of the practical applicability of that knowledge. To use the analogy above, biology has more in common with physics than with engineering.

The ulcer example may serve to illustrate this distinction. The isolation and identification of the H-substance in the 1920s was also in those days regarded as an important breakthrough, in the sense that it inspired further research and carried the promise of further acquisition of new knowledge. However, it was primarily a biological breakthrough, as the specific medical consequences could not be envisaged at the time. On the other hand, the results of the clinical trials of the new drug represented a major medical breakthrough, while their importance in purely biological terms was perhaps limited.

The distinction between biological and medical research is by no means clear. I said that the isolation of the H-substance was primarily a biological breakthrough, but it has been found repeatedly that new biological knowledge can usually be exploited for medical purposes. Therefore, even clinicians often engage themselves in basic research, and new biological discoveries are sometimes found worthy of a medical Nobel prize in spite of the fact that they have not yet been applied for practical purposes.

We also sometimes see medical breakthroughs which are not the result of new biological knowledge. For instance, the prevention of scurvy by fresh fruit, the prevention of smallpox by vaccination, the introduction of aseptic techniques, and the discovery of the effect of *Digitalis purpurea* are important landmarks in the history of medicine despite the fact that the mechanisms of action were quite unknown at the time. Even today we use many treatments whose mode of action is unknown, and most drugs used for the treatment of hypertension were introduced on the basis of hypotheses which later proved to be wrong. Such examples of medical progress can hardly be called biological breakthroughs, but they pose important biological problems which invite biological research. So also in these cases medicine and biology go hand in hand.

This presentation must not lead to the misconception that medicine is no more than applied biology, i.e. applied natural science. This is not the place to discuss the concepts of health and disease in any detail, but it must be mentioned that most of those who have attempted an analysis agree that these concepts cannot be defined in objective, biological terms. Health and disease are partly biological concepts, but they also depend on subjective feelings and on individual and cultural norms. This means that medicine belongs both to natural science and to the humanities or, to put it differently, that scientific medicine serves a purpose which, at least to some extent, reaches beyond natural science.

The ulcer example, which is representative of most medical research, leaves no doubt that scientific medicine is an important tool, but at the same time it illustrates the current paradigm of medical thinking which may be labelled biological reductionism. According to this mode of thought disease is conceived as an objective, measurable deviation from normal, biological structure and function, and it is seen as the purpose of medicine to correct these deviations. The subjective symptoms and the subjective feelings of illness are regarded as secondary phenomena and not as the primary constituents of the concept of illness. In the example, the scientific process started when French doctors decided to define a disease entity on the basis of an anatomical lesion and it reached its (temporary) conclusion when we knew so much about the functions of the human organism that we were able to eliminate this lesion. Within this paradigm the aim of medicine is defined in objective, scientific terms and it is simply taken for granted that progress towards such an aim is identical with progress towards the true aim of medicine. In the case of peptic ulcer disease this is undoubtedly true as most ulcer patients will feel that their health is improved when their ulcer has healed, but in other cases, which I shall discuss later, it is extremely important to remember that there is a difference between scientific progress and medical progress.

3 DIFFERENT TYPES OF MEDICAL RESEARCH

The road to success as exemplified by the development of an effective ulcer treatment is extremely complex as many different kinds of research are involved. Of course, we should like to see a steady flow of knowledge from one category of research workers to the next, but more often the process is characterized by standstills, bottlenecks and unnecessary duplication of research. In order to analyse some of these problems, we shall have to discuss which types of research are involved, and one of the best places to obtain a general impression of all the work which is going on within a particular field is a medical congress.

In 1976 a few thousand gastroenterologists, i.e. research workers and clinicians interested in the diseases of the stomach and gut, met at an international congress in Budapest. In all, 844 original papers were presented and, according to an analysis made by two colleagues and myself, they could be classified as shown in Table 1.

Table 1 *Analysis of research papers presented at congress in Budapest in 1976*

1 *Descriptive clinical studies*	120 (14%)
2 *Epidemiological studies*	21 (2%)
3 *Laboratory research*	
Animal studies	113 (13%)
Studies on normal human subjects	49 (6%)
Studies on patients	194 (23%)
Basic therapeutic research	43 (5%)
	Total 399 (47%)
4 *Assessment of diagnostic and therapeutic methods*	
Uncontrolled descriptive studies	200 (24%)
Controlled studies	36 (4%)
Studies of side-effects of drugs and complications of diagnostic tests	15 (2%)
	Total 251 (30%)
5 *Other studies*	

Of course, I do not suggest that these figures are representative of medicine as a whole. The large number of clinical papers (categories 1 and 4) reflect that the topic of the congress was a clinical one, and there are other congresses on theoretical topics where almost all papers are laboratory-orientated, but the results are in sound agreement with the distribution of papers at more recent clinical congresses and in many medical journals, and they may serve as a basis for discussion. Therefore, we shall briefly consider the importance and quantity of each of these types of research.

Descriptive clinical studies are important as they help to identify those practical problems which have to be solved. As mentioned earlier, numerous clinicians have been engaged in studying the clinical picture, complications and course of peptic ulcer disease, and the results of such studies provide much of the contents of our textbooks. In this way they help clinicians to make an exact diagnosis and to assess the prognosis in the individual case. Papers of this kind are common, but at the same time they are not held in high esteem by the medical profession. The contents of many such papers are trivial, and only a few descriptive studies are remembered as landmarks in the history of medicine.

Epidemiological studies in which research worker correlate the occurrence of different diseases to environmental factors are perhaps attracting an increasing interest. Nevertheless, only a small minority of medical research workers are interested in this kind of laborious research, and also in this respect the ulcer example is quite typical. We know that environmental factors play an important part in the development of this disease, but they have not yet been identified. We do not know if the problem can be solved, but the effort to solve it has been small. This is one of the bottlenecks in the research process, and I shall return to the problem later.

The vast majority of scientifically minded doctors are engaged in the exploration of the structure and functions of the human body in health and disease. They may simply be seeking new knowledge (i.e. they are doing biological research) or they may be trying to solve a particular disease problem. Such studies are usually the result of *laboratory research*, and at the congress in question, where clinicians dominated the picture, they accounted for 47 per cent of the papers. The popularity of this type of research is to a large extent the logical consequence of the current paradigm of medicine, i.e. the way in which doctors think, but other factors which favour laboratory studies will be mentioned below.

The results of laboratory research must be tested at the bedside before they can be exploited in everyday medicine, and for that reason clinicians do a large number of studies in which they *assess diagnostic and therapeutic methods*. However, as we shall see later, this is also one of the problem areas of medical research. It was mentioned that the effect of the new drug for ulcer treatment was tested by controlled trials, but in general there is still a lack of high-quality clinical trials. Many clinical trials are still of poor quality and at the Budapest congress the majority of assessments were little more than accounts of the uncontrolled experience of individual clinicians.

4 A CRITICAL ATTITUDE

Until a few decades ago most people inside and outside the medical profession probably agreed that the development of modern medicine was a story of success. From, say, 1850 to 1950 medical scientists could report new important advances every few years. I need mention only a few of these: discovery of bacteria and introduction of vaccination; development of endocrinology and effective treatment of diabetes mellitus and other endocrine diseases; discovery of antibiotics; effective treatment of pernicious anaemia; development of modern anaesthesiology and new surgical techniques. Of course, it is impossible to say how much of the improvement in the state of health of the populations in western Europe and North

America was due to medical progress and how much was due to better living conditions, but there can be little doubt that the contribution of medical science was important.

Since then, however, the attitude to medicine has become more critical, and there are several reasons for this. Breakthroughs are less frequent and their effect on the general state of health in the population is less conspicuous. Further, the problems of health in modern western society differ greatly from those which doctors had to solve in the last century and at the beginning of this century. In those days most hospital patients were young or middle-aged, and many suffered from infectious diseases, whereas, for instance in Scandinavia today, most medical hospital patients are old people with degenerative or malignant diseases. In addition, we see an increasing number of patients who suffer from the effects of alcoholism and others with inexplicable somatic symptoms which, in our ignorance, we label as psychosomatic. There is a general feeling that such problems are much more difficult to solve within the conventional framework of medical thinking.

However, the criticism of modern medicine must not be taken too far. Also in recent years we have seen important medical advances like the development of modern immunology, the introduction of transplant surgery, antenatal examination for chromosome abnormalities and the development of new diagnostic imaging techniques (ultrasound examination and computer-aided tomography). There is little doubt that medical thinking is changing in these years and that, consequently, our criteria for assessing medical progress will change, but at the same time there seems to be no need for a nihilistic attitude to modern medical science. In the following sections I shall criticize the present state of medical science on a number of points, but I shall avoid extreme, unwarranted criticism.

5 THE QUALITY AND QUANTITY OF MEDICAL RESEARCH

The format of medical scientific papers has been standardized to a remarkable extent. The typical paper is quite short (three to five pages in print) and it is divided into the traditional four sections: 'introduction', 'materials and methods' (or 'patients and methods'), 'results' and 'discussion'. This structure fits hypothetico-deductive research, as the authors describe their hypotheses in the introduction, then report the results of their experiments or survey, and finally discuss these results in the light of the hypotheses. Sometimes, however, medical papers are mainly descriptive and then it is the purpose to generate rather than to test hypotheses.

Medical scientists are, of course, expected to quantify their results as precisely as possible but, in contrast, for instance, to physicists and chemists, they always have to interpret their findings against the background of biological variation. The acid production in different ulcer patients varies and the effect of different drugs on the acid production also varies. For this reason the results must usually be subjected to a thorough statistical analysis and it is difficult to find a report of original research which does not contain the results of significance tests and statistical measures like the standard error or the correlation coefficient. However, medical research also presents many other pitfalls, like the placebo effect when patients receive a new treatment, and inter-observer variation when different clinicians record the same finding. There is ample literature on biostatistics and clinical research methods in which such problems are discussed.

There is little doubt that the *intrinsic quality*, i.e. the quality from a methodological point of view, is much higher today than only two or three decades ago when biostatistical methods first gained acceptance by medical research workers, but there is still ample room for improvement. There are many papers of high quality, but at the same time even the most reputed medical journals contain numerous papers with serious deficiencies. The results may be biased due to a methodological error, the conclusions may not be a logical consequence of the results and, most frequently, the authors have used the wrong statistical tests or interpreted the statistical results incorrectly. Anybody who teaches research methods will agree that it is possible to find examples of most classical methodological errors by browsing in a few medical journals.

The *extrinsic quality* of medical papers, i.e. their originality and importance, also varies. Of course it is dangerous to make rash generalizations, but I think that most readers of medical journals will agree that too many papers are concerned with trivialities or represent unnecessary duplication of the work of others. It is also quite common that authors only interpret their own results and that they do not consider the results of others in order to reach a balanced view of the scientific problem in question.

These deficiencies must be seen against the background of the quantity of medical papers. The number of medical journals rose sharply in the 1960s and 1970s, and the largest indexing services, which cover 2500 to 3000 medical journals, index around 250 000 references annually. However, the number of published papers must be much larger.

If my contention is true that too much medical research is substandard, valuable resources are wasted and we must consider in which ways the standard can be improved. The following suggestions are by no means original, but they should be mentioned.

It is the strength and not the weakness of medical science that so much of the work is done by medical practitioners, as it helps to ensure that the research activity on all levels is directed at the current health problems in society. However, it is a problem that most research workers have not been taught research methods and biostatistics. In Denmark, we believe that it has had a considerable effect that a large number of junior doctors have participated over the years in six-day courses on these topics, and undoubtedly much greater effects could be obtained if everybody who planned to do medical research, either part-time or full-time, was expected to study research methods for just one or two months.

Another problem is the career motive. In most countries young doctors who aspire to a career either as a hospital specialist or as a university teacher are expected to produce a list of their publications when they apply for a position. However, to put it very bluntly, the length of that list often seems to be more important than the intrinsic and extrinsic quality of the individual papers. It might be conducive to quality if applicants for such positions were allowed to present only a limited number of papers from their total scientific production, and if these papers were assessed critically. This suggestion does not solve the problem that some scientific institutions may be required to produce a steady flow of publications in order to ensure the necessary economic support.

The quality might also be improved if the editors of medical journals required higher standards. Many editors rely on the assessments by referees who are selected because of their knowledge of that particular field and who are therefore well qualified to judge the extrinsic quality of a paper, but who may not possess sufficient knowledge of methodology and biostatistics.

6 THE TWO AIMS OF MEDICAL RESEARCH: PREVENTION AND TREATMENT

It has been mentioned above that only a small part of the collective medical research effort is devoted to epidemiological studies and other studies of the influence of environmental factors on the development of disease. This is an important problem from the point of view of the advancement of medicine, and in order to explain the implications it is necessary to consider briefly the disease process in the individual person. The *etiology* of disease is the sum of those factors which elicit the disease process, and a distinction must be made between genetic and environmental factors. In most cases it must be imagined that genetic and environmental factors interact, but from a practical point of view the environmental factors are of greater importance, as we can change the environment but not the genetic constitution. The *pathogenesis* of disease, on the other hand, is the sum of those structural and functional changes which take place inside the human body as the result of the influence of the etiological factors.

When I state that there is a lack of epidemiological research, I really mean that there is a lack of etiological research. The majority of medical research workers are busy exploring physiological, biochemical and immunological mechanisms in the laboratory, which means that they are concentrating mainly on pathogenetic research. The difference between these types of research is important, as etiological research may teach us to prevent the onset of disease, whereas pathogenetic research in most cases only teaches us to interrupt the disease process and to treat a person who is already ill.

Admittedly, the distinction is not clear in the case of some diseases (for example, infections where the bacteria play both an etiological and a pathogenetic role), but it is essential when we are dealing with those malignant, degenerative and other chronic diseases which dominate the picture today. It is curious that most people outside the medical profession are unaware of the fact that doctors know next to nothing about the external factors which cause most of those diseases which they treat, apart from infectious and allergic diseases.

There are a number of reasons for this state of affairs. First, medical research is to a large extent guided by those practical problems which doctors are supposed to solve, and medical practitioners, especially doctors at university hospitals, spend most of their time treating sick people. Usually, they are not actively engaged in the prevention of disease.

Secondly, the facilities for pathogenetic research are often excellent both at university institutes and at teaching hospitals, and smaller projects may not even need funding from external sources. Large epidemiological surveys and studies of the occurrence of disease in different environments, such as different industrial settings, present much bigger organizational problems and invariably require external funding.

Thirdly, the career motive undoubtedly plays a part. It is understandable that junior research workers are more attracted by small laboratory projects which can be concluded quickly than by time-consuming surveys in which they are only a member of a team.

Finally, there is the subtle point that the disease classification which serves to define medical problems is mainly therapeutically orientated. Most diseases are defined on the pathogenetic level (for instance, peptic ulcer or hypertension) and it is quite likely that the same disease in different persons is determined by different constellations of environmental factors. It has, for instance, proved possible to determine a considerable number of so-called risk factors in studies of arteriosclerotic heart disease, but in each individual patient only some of these factors are present. In order to develop preventive medicine we must do laborious prospective studies of people who are exposed to a certain environmental factor (for example, eating certain foods or inhaling certain substances), and it is quite possible that we shall find that a single factor promotes the development of a number of diseases. Cigarette smoking, for instance, promotes the development of diseases as different as lung cancer, bladder cancer, heart disease and chronic lung disease.

Hopefully, we shall see an increasing amount of etiological research in the years to come in conjunction with the growing awareness of environmental problems in society in general.

7 THE APPLICATION OF NEW KNOWLEDGE FOR CLINICAL PURPOSES

In the ulcer example it was mentioned that the new antihistamine was tested by controlled clinical trials, but the survey from the Budapest congress showed that this type of research represents only a minute part of the total research effort in medicine. Of course, it is not possible to state what percentage of medical papers ought to be controlled trials in order to ensure optimum utilization of new knowledge generated in the laboratory, but there can be little doubt that this is another problem area in clinical research.

The situation has improved considerably during the last few decades, as the health authorities in many countries demand the evidence of controlled trials before they permit the registration of a new drug. In such a trial the patients are allocated at random to two groups, one of which receives the new drug, while the other receives the best current treatment (which may well be inactive tablets). In order to avoid psychological bias, the trial is often 'double-blind' so that neither patients nor doctors know which patients receive which treatment. Today there are a considerable number of clinical centres which conduct research of this kind, and doctors who do clinical trials may also receive the advice of experts from the pharmaceutical companies.

However, the situation is very complex. The number of good drug trials is certainly on the increase, but unfortunately only a few new drugs represent progress from a medical point of view. Many trials, perhaps the vast majority, concern drugs which resemble those which are already in current use and which have been developed only for commercial reasons. It is also natural, from a commercial point of view, that the drug industry is particularly interested in the development of new drugs for common chronic conditions, like hypertension, heart disease, peptic ulcer, rheumatic conditions and nervous disorders. In these areas we see a steady but quite unnecessary increase in the number of new drugs, while the research effort in other areas is limited.

There is also a great lack of controlled trials when this type of research is not promoted by the requirements of health authorities and the interests of the drug industry. We ought to test many of the drugs which were introduced long before the present requirements were enforced, and we ought to test a large number of other treatments, like surgical techniques and physiotherapy. In those areas relatively little is done, and there are several reasons for that.

First, controlled trials are laborious, and many clinicians still feel that they are not really necessary. They believe that they can predict the result of a treatment by deductions from their theoretical knowledge of disease mechanisms and that they can trust their uncontrolled experience at the bedside. However, there are numerous examples that treatments, which doctors thought were effective, were later found by controlled trials to be quite ineffective or even harmful.

Secondly, the general public views such trials with some misgivings. It is generally assumed by everybody that medicine must progress, but it is felt unacceptable that doctors cannot prove by indirect means, such as by animal experiments, that some treatment has the desired effect.

It is obvious that patients must be fully informed of the trials in which they participate and that the research protocol must be approved by an independent ethical committee. However, it has not yet proved possible to convince everybody of the necessity of controlled trials, and discussions in the press in some countries may have made clinicians even more wary of engaging in this type of research.

Thirdly, controlled trials, like epidemiological studies, are typically conducted by a large team, and many junior doctors prefer small laboratory projects which they can do on their own.

New diagnostic techniques ought, in principle, to be tested as critically as new therapeutic methods, but this is very rarely done. I shall not discuss this complex topic in any detail, but just mention that new technological advances often find a place in the clinical routine in spite of the fact that it has not been proved that it serves the patients' best interests. Sometimes it has not been proved empirically that the new technique improves diagnostic accuracy, and sometimes it has not been ascertained that improved diagnostic accuracy has any influence on the outcome of the patients' illness.

8 BEYOND NATURAL SCIENCE

As mentioned above, medical research workers traditionally regard medicine as applied biological science. This view may have been acceptable in former days when most medical treatment was ineffective and when doctors simply hungered for more knowledge, but today we have powerful diagnostic techniques and treatments at our disposal, and we can no longer ignore the fact that both medical research and medical practice involve value judgements.

As an example, some clinical trials show that combinations of cytotoxic drugs with unpleasant side effects may slow down the disease process in patients with malignant diseases. Sometimes they cannot cure the patient but they can prolong his life. In such cases the objective scientific facts which are the results of the trials cannot be used for decisions at the bedside, unless the doctor is willing to balance the value of survival for a limited period of time against the patient's quality of life.

People outside the medical profession are aware of this problem, and scientific medicine is regarded with a mixture of awe and fear. In Scandinavia, for instance, old people sometimes express the fear that they may not be allowed to end their lives in a dignified way and to die a natural death if they are admitted to hospital, but that they will be kept alive for as long as technically possible, regardless of their quality of life.

Other difficult problems arise from the fact that the part of the economic resources which a society can spend on health problems is limited, and it is necessary on all levels to think in terms of priorities. Members of a medical research council may well have to decide whether they should support a study on the causes of alcoholism or a study on immunological reactions after transplant surgery; hospital planners may have to choose between buying a CT-scanner or extending a geriatric ward; and the individual clinician who reads about controlled trials of new drugs in ulcer disease must keep in mind that these drugs are very expensive, and that their use in all cases of ulcer disease will require that a large sum of money is saved somewhere else in the health service.

All these problems have been forced upon a medical profession who was taught to think almost exclusively in scientific terms, and the result has been an increased interest in medical ethics. The analysis of value judgements has become a medical discipline in its own right, and although this development can hardly be called a scientific breakthrough, it could be heralding a revolutionary change in the paradigm of medical thinking.

In the no-man's land between the science and the art of medicine a new field of medical research is developing, which is based on clinical decision theory, i.e. the application of statistical decision theory to medical decision-making. Those who are engaged in this type of research take into account both empirical scientific knowledge (i.e. the probability of different events) and value judgements (i.e. the utility of those events), and in this way they are trying to bridge the gap between medical science and medical practice. Because of the subjective element such studies cannot be assessed according to the same criteria as empirical clinical studies.

9 ASSESSMENT OF PROGRESS IN PSYCHIATRY

Up to now I have discussed only so-called somatic medicine, and it would be wrong to ignore completely the problems of psychiatric research which is much more heterogeneous.

Most – but not all – psychiatrists will agree that the development of psychoanalysis represented a major breakthrough in the history of psychiatry. Freud regarded himself as an empiricist, but nevertheless the psychoanalytical tradition in psychiatry is much more closely related to the phenomenological tradition in continental philosophy than to the empiricist tradition of natural science. The assessment of progress in this area is very different from the assessment of progress in scientific medicine as it depends more on sympathetic insight than on objective facts.

Other psychiatrists regard the psychoanalytical tradition as unscientific (which, of course, is correct if science is equated with objective, natural science), and instead they view mental disease from other angles. They may, for instance, belong to the behaviourist school, which means that they regard normal and abnormal human behaviour as the result of complex reflex patterns which to some extent can be changed by reconditioning. The critics may also belong to the school of biological psychiatry, which means that they hope to solve the problems of mental disease by exploring the underlying neurophysiological mechanisms. Research of this kind, which is little different from biological research in somatic medicine, has led to effective drug treatment of some mental disorders as well as to widespread abuse of a large number of tranquilizers.

Because of the heterogeneity of psychiatric research, it is at present impossible to establish criteria for the assessment of progress which would be acceptable to all psychiatrists.

10 CONCLUDING REMARKS

Progress in medicine depends on close co-operation with scientists and scholars from a variety of disciplines.

Traditionally, medicine has been regarded as applied biology, and there is no sharp distinction between scientific medicine and disciplines like general biology, biochemistry and physiology. Medical technology is the result of progress in a large number of technological areas – for example, electronics, fibreoptics, ultrasonography and laser technology – and these technologies depend on progress in, for instance, pure physics. Medical research, both in the laboratory and at the bedside, makes use of biostatistics, and therefore there is also a link to mathematics. The different modes of thought in psychiatry reflect different traditions in psychology, and psychoanalysis is more closely related to the humanities than to the scientific disciplines. Further, epidemiological medicine is related to sociology, and medical ethics are both a medical and a philosophical discipline.

The only feature which distinguishes for certain medical activity from all other scientific and academic activities is its aim, i.e. the promotion of health. Therefore, medical progress also includes progress in all the related disciplines which serve this unique aim.

Comment

H.-R. DUNCKER

In his very clear example of the historical development of the medical treatment of peptic ulcers, Dr Wulff points out that the present success in this treatment is the result of progress in very different disciplines, from biology, physiology and pharmacology to internal medicine and surgery. Both biological discoveries and developments in medical knowledge have been responsible for the breakthroughs that make up progress in medicine. Since 1950, however, the attitude towards medicine has become more critical. Breakthroughs are less frequent and general change in the state of health in western populations is less conspicuous. A large proportion of hospital patients are geriatrics with degenerative or malignant diseases, while others in increasing numbers suffer from alcoholism or so-called psychosomatic diseases. These problems are much more difficult to solve within the conventional framework of medical thinking.

Beyond these studies based on the natural sciences, Dr Wulff sees an increasing necessity for complex value judgement of the far-reaching medical treatments and techniques employed today, which often have such unpleasant side-effects that the patient's quality of life is drastically reduced during his limited period of survival. Dr Wulff summarizes that progress in medicine depends on close co-operation with scientists of a variety of disciplines from the traditional basic applied biology and medical technology to psychology, psychoanalysis and sociology. In this way, Dr Wulff defines the complex situation of medicine, which, because of its important role in society, has to integrate for each generation the newly developed knowledge from those disciplines. The increasing knowledge of functional relationships in the early nineteenth century, not only in chemistry and physics but also in the biological and medical sciences, attracted most of the young scientists of that time, who spent all their efforts in the development of these experimental disciplines. The results provided the possibility of altering natural processes: the utilization of material and energy resources was responsible for the enormous increase in living standards, and insight into the functional processes of the organism enabled medicine to develop for the first time effective treatments capable of conquering illness and prolonging life expectancy.

This way of thinking was so dominant that our knowledge of structural order and relationships was largely neglected. Since 1950, as Dr Wulff mentions, this lack has become conscious again, but without sufficient knowledge of how to change the basic situation. Therefore, I want to draw attention to a phenomenon which, in my opinion, is a major problem in both medicine and biology. Even today, the general thinking of medical scientists on the nature and function of the human organism suffers from the fact that physics is the model science and its doctrines of functional relationships are the generally applied way of thinking, even used in the humanities. The specific nature of organisms, societies and civilizations with their inherent characteristics, going far beyond the functional properties, are seen only as unwanted side-effects. The present situation of medicine is not only characterized by a strong reductionism towards functional biology, but biology itself also suffers from its own reductionism to physics. A proper integrated picture of the whole organism is lacking in biology today and thus medicine also lacks a modern view of the human organism in its entirety.

As far as we are capable of understanding whole organisms, they are historically developed, complex structural–functional systems, with a multiplicity of interlinked, hierarchically organized levels of functional systems, each with its own set of functional abilities and qualities. Within the limits of our present knowledge, the basic principles of molecular biology, biochemistry and biophysics are governed by the laws of chemistry and physics, outside which life on Earth does not exist. These basic principles are interlinked to more complex structural–functional systems that possess higher regulatory abilities and create new functional potentialities in cellular and developmental physiology. The more complex systems give rise to differentiated organisms with a number of different tissues and organic systems. The resulting multiplicity of organismic structure and function is adapted to a broad range of ecological conditions. During evolution, new levels of functional and regulatory systems were derived primarily not by the origin of new elements, but rather by the functional linkage and interaction of previously existing, unrelated systems, endowing the organism with new functional capacities. In this way, the human organism also acquired its specific structural and functional capacities by a long historical, phylogenetic development.

All single functional relationships, or the interdependence of single components of functional processes in organisms, regardless of the level at which they act (from the biochemical to the cellular and organic levels up to higher brain functions), can be investigated experimentally. But this is possible only after the functional system or process, of which the functional interrelationship investigated is a part, has been discovered as a whole. The enormous progress in medicine in the 100 years preceding 1950 was a result of increasingly detailed analyses of known functional systems gained from the experimental disciplines. From these successful discoveries, a vast number of fascinating facts have been gathered, which have enabled medical scientists to alter the functions of the human body and develop highly effective medical treatment of a large number of illnesses. This is the basis of modern medicine and its success, but also of its present problems.

These problems result from the fact that scientific development of the experimental biological and medical disciplines in the last 100 years has ignored their historic dependency, and even worse, the historical nature of the organism. All experimental analysis is based on a previous, more or less precise knowledge of complex structural–functional systems, from which single functional processes are selected to be analysed experimentally. Knowledge of the complex, hierarchically interlinked systems which make up the organism cannot be acquired by experiments, but only by comparative investigations.

This is the old way of thinking in analogies, to apply a known picture or pattern of thinking, gathered perhaps from quite different fields, to newly acquired observations of structures or functional relationships. Thus, a previously unknown or inconceivable system or element of an organism can be described for the first time and an initial concept can be formulated to cope with this new information and integrate it into the already existing body of knowledge.

This initial concept of the general design of the newly observed structure or functional system can be tested and further differentiated by comparing related structures and functions in a broad range of different organisms. Thus, a refinement of the concept occurs by going back and forth between the direct observation of the components of a new system and the scientific formulation upon which it is based. In this hermeneutic spiral, the concept comes nearer to the exact description of the system. To discover an unknown complex structural and functional system in the organization of an organism, it is necessary to take more details into account. Therefore, the best way is to compare organisms with different historical backgrounds, functional adaptations and variations in body size. In this way, the general structure of the system can be elaborated and differentiated into characteristic specializations, typical only for a single species. The broader the basis for comparison, the more relevant the results. Single species cannot be compared. One major problem of current medical research is the use of laboratory animals which are, according to general thinking, only small, manageable models of human organisms. Their specific organization, quite different from human organization in many functional systems, is not recognized.

Using this methodological approach, the comparative investigations of biological and medical scientists at the beginning of the nineteenth century formed the basis upon which the structure and function of organisms was elaborated. From these comparative studies we attained a broad knowledge of living organisms, their internal structure and their relationship to one another. The 'natural system' of the organisms, which is still relevant today, was developed from these studies and later interpreted as a phylogenetic system. These investigations, founded on other comparative investigations of culture and linguistics at that time, presented our basic knowledge of the functional systems of organisms, upon which the relevant experimental analyses depended as a source of fundamental knowledge.

In a critical examination we have to consider that medicine, depending for its success in this century mainly on the results of the experimental sciences, has totally lost the comparative aspect of organisms and thereby the knowledge of the historical nature of the human organism. Organisms, including the human organism, are not simply the sum of all functions, but they are historically developed, complex structural–functional systems, in which all specific structural and functional features are the result of discrete historical changes, preserved by the genetic code, having been selected as successful alterations in evolution. In this way, our body represents the long history of vertebrates from sessile filter feeders to lung-breathing fishes, land-conquering tetrapods, homoiothermic mammals, whose viviparity enabled a prolonged gestational period, a necessity for higher brain development. Thus, the basis for all learning and communication was established, upon which the evolution of higher social primates and lastly language-using human beings depends, including a long childhood and cultural tradition. Thus, our organism is imprinted in its structure and function by its history, and we can demonstrate the remnants of these historical steps in the structure of our body.

This historical nature of the human organism cannot be discovered experimentally, but only comparatively as in the historical sciences. During the dominance of the experimental disciplines, the further elaboration of the hierarchically structured and interlinked functional systems stagnated: these systems are known only incompletely. The functional systems that regulate the growth and size of the organs and the body are widely unknown. We are just starting to get an insight into morphogenetic processes, which not only regulate the embryonic development of the body, but which seem to be responsible for the life-long maintenance of the size and relationship between tissues, organs and the body. Many chronic degenerative diseases of older people seem to be related to slight aberrations in these processes. An unknown number of such functional and structural interrelationships govern the growth and function of our body, which are determined during our long phylogenetic history.

If we want to obtain a better understanding of our body and its functions we have to respect these historically determined interrelationships. If we want to overcome the present 'repair medicine' we have to realize that many functions of our body are not designed optimally from a technical point of view, but are historically determined in a certain manner. The functioning of organs is designed for a limited lifespan, which guarantees during youth high performance and reproduction under the given conditions of social life. With increasing age and reduced performance, the lifestyle has to be changed and adjusted to the altered conditions and possibilities. For old, retired people, the way of life brings further changes and new goals. Apart from all social and psychological aspects, this fact results from the basic premise that the limitation of our lifespan is not a technical defect of our body that can be repaired by modern medicine, but an inborn constituent of the life of the organism and each individual life. To respect these basic conditions of our life will help to adjust medical treatment to the nature of our body and its requirements at different ages. It will help to create more humane medicine, which can be characterized by the tenet 'not always to do everything which is possible, but to undertake those treatments that are adequate to the situation and age of the patient'.

This change in the goal-directed activity of medicine with an altered picture of the human organism and its health requires a change in the orientation of medical science and education. The above-mentioned way of understanding human organisms as historical beings requires the reincorporation of the comparative aspect into the training of medical students, as was usually done up to the beginning of this century. In this way, they will get at least a feeling of the historical nature of the human organism, which incorporates a large number of boundary conditions that cannot be simply ignored by medicine, including a natural, dignified way of dying.

For the fruitful development of medicine, both types of analysis – first, the experimental, causal–analytical investigation of functional processes and interdependencies, and secondly, the comparative, hermeneutical pattern analysis of complex structure–function systems – have to be performed complementarily. There is a methodological problem today because the hermeneutical pattern analysis, common in cultural sciences and humanities, is in biological and medical sciences most frequently rejected as 'unscientific'. From this point of view, it has to be stated that the transition between the exact, experimental sciences, like physics and chemistry, and the comparative, historiographic and hermeneutical disciplines, like cultural and language sciences, lies at the centre of the biological and medical disciplines because of the nature of organisms as historical beings. Their functions are governed by the exact laws of the natural sciences, but the arrangement of their complex functional systems depends on a large number of historically unique events, which are preserved in the genetic code and selected by survival. Therefore both these methods of analysing reality have to be applied in a complementary way, in order to emerge from the present one-sided situation of the biological and medical sciences.

This comparative view of medical science has another fundamental aspect for the development of a realistic, more holistic picture of the human organism. A crucial point for the development of such an overall picture is the situation that a separate discipline has been developed for the analysis of each level of the functional systems in the organism, each with its own methodology, terminology, recognition and theories. The basic functional levels of the organisms are analysed by the experimental disciplines of biochemistry and molecular biology. Physiology and pharmacology with their various fields start with experimental analysis, but approaching higher functional levels they advance to comparative investigation, best documented in the modern development of comparative physiology today. Morphology, psychology and sociology work mainly with comparative methods. The integration of the different levels in a holistic picture of the organism needs comparative elaboration of the general pattern structure of the functional systems at the different levels. Only then is it possible to demonstrate, in hierarchically organized systems, which structures and functions of an underlying level determine the design and capabilities of the next higher level, and where the new elements and qualities are to be found. This integration of very differentiated knowledge from various disciplines (from the basic constituent chemical and physical functions to highest brain functions) can only be performed by a comparative approach with its generalized patterns at the different levels of organization.

In this way, a more appropriate picture can be elaborated of organisms, especially the human organism, which may help us to understand more completely the complex interactions in our body and its limitations. But human beings are more than biological organisms. During the phylogeny of higher mammals, interactions between the organisms and the surrounding world lead not only to more highly evolved primates, but also to the generation of different forms of social life. These societies developed systems of communication, and the most highly evolved acquired traditions. By the development and use of language and social learning, these traditions were the basis for the historical development of the human social structure and social knowledge that make up our civilization. The biological systems that are responsible for the ontogeny, structure and function of our brain, and its interactions with our body, are the limiting conditions, the realm within which all our social, psychological and intellectual activities are manifest. These cultural activities created their own standards with newly acquired qualities, which, however, cannot be derived from the biological systems, but which are significantly influenced by the human organism who produces them. These complicated relationships are often neglected.

If we are looking towards the future development of medicine, it is necessary not only to develop a more organismic view of human beings, but also to observe how the human organism serves the biological structures and functions for the development of our social, psychological, intellectual and cultural activities. Further, it is necessary to elaborate the mode in which our thinking and feelings and our social interactions reflect back on the organism and its structures and functions. To develop such a holistic view of human beings it is not sufficient to add to a 'biologically reduced' medicine new disciplines such as psychology and sociology. The incorporation of these aspects will be fruitful only when a picture of the organism as a historical, hierarchically organized, complex structural–functional system with its inherent continuous interaction with psychological, social and cultural aspects has been reintroduced into the general thinking of medical scientists and practitioners.

With such an organismic view of human beings, psychosomatic medicine will no longer be such an isolated discipline, but an integral component of all medical activities. With such a modern, renewed organismic concept of human beings, a number of current medical problems will be solved much more readily than seems possible at the moment. Some of the so-called ethical problems in geriatric or intensive care medicine will be solved as a result of a more adequate organismic, psychosomatic and cultural view of the human organism. Such a view serves as a better basis for the remaining true, ethical problems. The social, psychological and cultural relationships of a patient, in addition to the biological and medical problems of his illness, have to be incorporated into the whole picture of the patient, and this is only possible in the realm of a complex value judgement, which Dr Wulff described as an increasing necessity. New ways have to be explored for such a value judgement, in which must be integrated not only the individual constituents of the patient, but also the subjective situation of the medical practitioners engaged in these decisions. This is a difficult task for the future development of medicine.

I want to conclude this commentary on progress in medicine by asking a question that has already been raised in the discussion: why is there a lack of theoretical medicine as a scientific discipline? In the last few decades, overwhelming amounts of new experimental data and totally new insights into functional processes have been elaborated and yet no final picture has emerged. In a number of research fields, knowledge has undergone a rapid evolution. Within the premises of these expanding experimental sciences, it is an unspoken expectation to attain a final generalized picture after having solved all the detailed questions. Unfortunately, this is generally an erroneous hope. The lack of theoretical medicine as a concrete scientific discipline is mainly the result of the lack of a theory of organisms. Theoretical medicine would have to incorporate into its general theory of organisms the specific human conditions that arise from human, social, psychological and cultural activities. Such a theory of organisms as the foundation for theoretical medicine will only be possible on the basis of a theory of the complex structural–functional systems in organisms. It can be elaborated not by experimental investigations, but only by general comparative analyses founded on hermeneutical methodology. In the rejection of this basic scientific necessity by most of the current biological and medical scientists lies the major hindrance for the development of theoretical medicine. This situation is the strongest expression of reductionism in the thinking of today's scientists, which can only be overcome by the reintroduction of the comparative aspects explained above and appropriate methodological procedures into medical investigations and teaching. If we strive to change the present inadequate situation in medicine, in my opinion, only the comparative approach can help to alter general reductionistic thinking to a more organism-oriented and consequently a more humanistic general attitude.

Scientific advancement in sociology

R. BOUDON

SOCIOLOGY is a discipline certainly characterized by a greater heterogeneity of its objects and of its methods than other social sciences, such as economics or demography. To some extent it is a *residual* discipline, in the sense that it deals with the problems which are not taken in charge by the 'special' social sciences. The singularity of sociology is revealed by the organization of sociology studies: basic sociology courses often take the form of a discussion of the work of the so-called founding fathers (Pareto, Weber, Durkheim, etc.), rather than of the presentation of a well-structured and articulated substance or formal body of knowledge.

Given the diversity of sociological production, in the past and still today, sociology is often considered as a discipline on its own, following particular rules with one foot only in science and characterized by a mode of thinking located halfway between philosophy and science.

Although this sceptical attitude contains some truth, it is more useful, it seems to me, to take a critical or relativistic view with regard to sociological production, i.e. to distinguish and identify the various types of sociological production. Such distinctions are indispensable in approaching the problem of the identification of scientific advancement in sociology.

I SOCIOLOGY CAN ALSO BE A 'HARD' SCIENCE

The first point to stress is that many sociological theories can be classified as belonging to 'hard' science in the sense that their confirmation or refutation follow, rather accurately, the processes described by the neo-Popperian epistemics, i.e. the classical Popperian principles as revised by such authors as Lakatos. In other words, these theories explain with parsimony a number of facts in such a way that the explanation can be rationally discussed and controlled.

I shall evoke briefly a few examples in this respect. The first is the well-known theory of Max Weber, according to which Calvinism would have played a crucial role in the development of capitalism in the sixteenth and seventeenth centuries. This classical theory attempts to relate a number of facts: that capitalism appeared to be more at home in protestant countries, that an impressive number of capitalist entrepreneurs were Calvinist, that even in the Lutherian countries such as Sweden or Denmark, the businessmen were Calvinist, etc. From these facts, Weber drew the impression that something in Calvinism should explain the fact that entrepreneurs were 'more than proportionally' Calvinist, and selected 'naturally' the feature which distinguished Calvinism from Lutheranism: the predestination dogma. The problem was then to explain the effect of this dogma. Weber's ingenious explanation – which is complicated and which I summarize in a very rough fashion – is that, as in business failure and success can be immediately and unambiguously identified, the businessmen had a way of recognizing whether the grace of God was with them or not. In other words, the belief in predestination should incite 'worldly ascetism', and generate a preference for accumulation rather than consumption, and for achievement rather than hedonism. Moreover, Calvinism insisted on the legitimacy of wordly activities which contributed to the glory of God, but the latter point would also be true of Lutheranism.

I do not want to insist more lengthily on Weber's original theory, which is well known; rather, I would like to insist on the further discussions of the theory, which was submitted to rational discussion at an early stage. Some of the arguments which were raised against it illustrate the famous Popperian falsification procedure:

(a) Entrepreneurs were on the whole more than proportionally Calvinists; but many of them were Jews or Catholics (in Cologne for instance).
(b) There had been capitalist entrepreneurs in the fifteenth century, before the Reformation, for example the Fuggers in Augsburg. Weber's view that fifteenth-century businessmen would have been different from those of the sixteenth century appears as *ad hoc*.
(c) Business was less flourishing at the time in Scotland than in England, although Calvinism was more strongly implanted in Scotland.

Thus, a first set of facts – the list is actually longer, and the three above facts are only examples – contradicts Weber's theory. Other facts, too, fail to be explained by this theory:

(a) The entrepreneurs active in Geneva were never born in Geneva, nor in Switzerland.
(b) The Calvinists active in Amsterdam were seldom born there.
(c) Milan, where Calvinism was strong, declined economically at the end of the sixteenth century.

And again the list can be made longer.

On the other hand, it was observed that the effectiveness granted to the predestination dogma in Weber's theory was to some extent *ad hoc*. In other words, it was completely 'understandable', in Weber's sense of the word, why belief in the predestination dogma would incite the typically capitalistic behaviour of accumulation.

Many scholars contributed to this discussion, the most important one being probably Trevor Roper, whose theory, while it keeps the core of Weber's original formulation, dissipates all the difficulties of Weber's theory and explains the facts which that theory did not explain. Briefly summarized, Trevor Roper's revised version of Weber's theory is the following:

(a) There were already entrepreneurs and businessmen in the fifteenth century.
(b) When Erasmism appeared, they felt 'naturally' inclined to adopt this ideology, which legitimated worldly activities.
(c) Calvinism is a radical form of Erasmism.
(d) When the Counter-Reformation appeared, business became more difficult in the countries under Spanish control because of the increasing weight of taxes, and also because the Counter-Reformation ideology was unfavourable to business.
(e) Consequently, many businessmen emigrated: they went from Antwerp to Amsterdam or Geneva, or to the Scandinavian kingdoms.

Trevor Roper's theory keeps the core argument of Weber but gives it a more comprehensible form: while it is difficult to accept the effectiveness of the predestination dogma, it is easy to understand (see (b) and (c) above) that the businessmen were attracted by Erasmism and then by the Calvinist international movement. On other hand, Trevor Roper's theory explains in a simple fashion all the facts that were either with difficulty compatible to the Weberian theory or were not explained by it.

Beyond doubt, Trevor Roper's theory is better than Weber's original one: it explains more; no contradictory facts may be opposed to it; and it is constituted by a set of statements, all of which are acceptable. The example is particularly interesting because it deals with ideological phenomena. Very often the view is held that such phenomena cannot be approached by traditional scientific procedures. The story of the Weber–Trevor Roper theory shows that, even with such phenomena, a rational discussion – similar to those considered normal in the 'hard' sciences and which the Popperians have described – can be established. Since this point is often hardly recognized, I shall mention two further examples.

First, the example of Hagen's theory of social change. Hagen was intrigued by the fact that some countries, notably Colombia at the beginning of the century, enjoyed a high rate of development, although, according to the prevailing theories, they should not have: markets in Colombia were local, the saving capacities very scarce, foreign investments non-existent, and overhead capital very poorly developed (there were almost no roads, for instance, and connections between the main cities were very precarious). In spite of these unfavourable conditions, Colombia developed rapidly, thanks to the apparition of an élite of entrepreneurs. This élite came more than proportionally from one province, Antioquia, which was a kind of backwater. I do not want to go into the details of Hagen's theory, suffice it to say that he explained the apparition of this élite convincingly by the different opportunity structures in Antioquia: it was impossible, or at least inadvisable, for members of the upper classes there to try to enter professional, cultural or political careers, and since there was no landed gentry, they were not incited either to buy land. In practice, business was the only way of status conservation and promotion open to them.

In this case too, we have a Popperian process of advancement; Colombia did not fit into the theories of development, and so Hagen looked for the peculiar factors responsible. The main factor was the existence of a marginal group which, given the structure of opportunities, was incited to develop business activities. Hagen noted that the same factor had applied in other cases: for instance, the Samurai in Japan, also a marginal group, played a more than proportional role in the development of that country.

My final example deals with the sociological theory of 'modernization'. A prevailing view in the 1960s and 1970s was that modernization had a kind of intrinsic tendency to progress simultaneously on all fronts. Thus, it was supposed that industrialization implied a nuclearization of family structure: as industrialization implies a certain level of mobility of the labour force, and as mobility is incompatible with the extended family structures prevailing in pre-industrial societies, industrialization should affect the family structures and contribute to substituting the nuclear model by the extended one. While the theory contains some part of truth, empirical research has shown that, in many cases, when industrial salaries are low, the industrial worker uses part of the salary to obtain from his orientation family goods and services at much lower prices than market ones; the family benefits from the surplus provided by the salary. On the whole, in such a case industrialization has the effect of reinforcing rather than weakening the relationship between the members of the extended family, so that the nuclearization announced by social change theorists is not actually observed.

All these examples show that sociology, more often than is sometimes believed, advances along the lines described by Popperian epistemics: precise questions are raised which are handled as puzzles, theories are built which attempt to explain the puzzles, new facts are detected that appear as contradictory to the prevailing theories, thus leading eventually to a correction of those theories, or to the construction of new ones.

Parenthetically, I would add that I have intentionally selected qualitative examples: those we have examined deal with 'hard' data, but of the qualitative type. It is often wrongly assumed that quantitative sociology, i.e. dealing with the analysis of quantitative data, is more 'scientific' than qualitative sociology. It can be, but it is not necessarily so.

Finally, it should be stressed that Popperian epistemics have continuously been present, at least implicity, in classical sociological research as well as in modern research. Tocqueville, Weber, Sombart and many others developed sociological theories of which the logical nature is not basically different from those built by the 'hard' sciences, and which were generalized, falsified or confirmed by procedures similar to those currently in use.

2 A PLURALITY OF PARADIGMS

So far, so good: sociology is rightly perceived, from the outside and from the inside as well, as a discipline with strong singularities. Not all sociological products and discussion follow Popperian epistemics. Beyond doubt, the epistemological consensus is weaker in the case of sociology than in other sciences, even social sciences.

Traditionally, there has been a lasting discussion among sociologists on the basic principles of sociological explanation. This discussion can be perceived in the methodological work of Weber and Durkheim for instance, who did not agree with one another, as well as in more modern works. Structuralists, for instance, do not agree with action theorists, and a leading German sociologist, Habermas, in a famous discussion, tried to show that sociology could not follow the same kind of rationality as other scientific disciplines. In this section, I will try to describe these classical and modern discussions. I will then raise the question of whether any progress or advance is noticeable. In contrast to the previous section, we will not deal with sociological theories proper, but with paradigms in the Kuhnian sense.

In the Durkheim tradition, sociology is meant to discover regularities, either in the form of time series (e.g. increase in the division of labour over time), or in the form of statistical correlations (e.g. correlations between rates of suicide and various indicators of similarity, such as rates of divorce). Individual motivations, according to Durkheim, cannot be scientifically observed and matter little. Thus, his sociological study of suicide pays no attention to motivations. It consists in studying the statistical variation of suicide over time, in various contexts, and in relating the rates of suicide to a number of other variables.

The German tradition, by contrast, as represented notably by Max Weber and Georg Simmel, stresses almost opposite principles. Georg

Simmel writes that the obstinacy with which social scientists look for regularities of the social life is the product of the old metaphysical creed according to which there could not be a valid knowledge, but of the universal and the necessary. For Simmel, as for Weber, discovering regularities cannot be the main task of sociology. And, in contrast to Durkheim, they insist on the fact that the typical objective of sociology is to 'explain' social phenomena as the outcome and the product of individual actions. So, a crucial moment of any sociological analysis is to 'understand' (*verstehen*) the motivations of the social actors. Durkheim treated Simmel's sociology as 'metaphysical' and had little sympathy for Weber's work.

Weber's and Simmel's paradigms are very close to what was later called 'methodological individualism', while Durkheim's representation of sociology makes it close to physics, as he conceived it at least. Pareto's paradigm is very close to Weber's. He also insisted on the fact that social phenomena should be conceived as the product of individual action. Like Weber, he stressed the point that the rational model of action as used by classical economists is particular rather than general, however.

The Weber–Durkheim opposition is still visible in modern sociology. Thus, the 'structuralists' insist on the fact that discovering structures, i.e. non-random combinations among social features, is the main task of sociology; most neo-Marxist sociologists for their part pay little attention to individual actors and to their motivations; they aim at discovering 'regularities in the social life', to use Simmel's expression.

A second opposition is more or less permanent in sociological discussions: while Weber (who agrees with Durkheim on this point) thought that sociology can be objective and follow the same procedures as any other discipline, other sociologists think that sociology, because it includes operations of interpretation which have no equivalent in the natural sciences, should follow a methodology of its own and cannot aim at objectivity in the sense of the natural sciences. Sociological analysis, according to these writers, would be closer to the interpretation of art works than to the explanation of natural processes as practised by the natural sciences.

The preceding paragraph is a very sketchy presentation of a recurring discussion that has taken many forms. Dilthey in the past and Habermas today are examples of the second type of attitude. Sometimes also the point is made that sociologists, as they are themselves social actors, cannot have a distanced and neutral view of social phenomena. This point was raised by Mannheim and is represented by such authors as Habermas and Touraine for instance, who follow certain suggestions of Marx in this respect. The problem is then to decide which viewpoint is better, since a neutral viewpoint is supposedly inaccessible. The answer often given in this respect is again close to Marx's: the 'best' viewpoint is the 'critical' one. In other words, the sociologist should disentangle the hidden mechanisms thanks to which the social structures are maintained, and the hidden forces which contribute to transforming these structures. In his recent work, Touraine introduces the notion of 'permanent sociology': the task of the sociologist would be to help the social movements, which are the main channels of social transformation, to develop and ripen. Though he uses other words, Habermas's views are similar.

The two oppositions which I have described are present in contemporary sociology at the international level. When taken in combination, they provide three types of basic sociological orientations or paradigms:

(a) The 'Weberian' one – sociology is not different from the other sciences; its task is to interpret a certain social phenomenon as the product of individual actions.
(b) The 'Durkheimian' one – sociology can be a 'hard' science, and should inquire into factual–observable regularities.
(c) The 'critical' one in the sense given to the adjective 'critical' by the 'Frankfurt School' – sociology is an interpretative science, it cannot aim at objectivity; it aims at disentangling the hidden interests and the hidden forces lying under the social structures.

As a matter of fact, one should divide this third paradigm into two components: the 'interpretative' one, which insists on the fact that social phenomena should be analysed with the same kind of methodology as used by art critics or novelists, and the 'critical' one, which also insists on the importance of interpretation, but wants to identify the latent forces responsible for the 'reproduction' or the 'production' of society.

Obviously, these three paradigms are hard to reconcile with one another. The fact that they are simultaneously present in contemporary sociology partly explains that the consensus typical of 'normal science' in the Kuhnian sense certainly does not prevail in our discipline.

I shall try to suggest very briefly – being conscious of the part of subjectivity in my position – that in reality the Weberian paradigm is the most fruitful, and that the objections addressed to this paradigm often rest on misunderstandings.

Point 1. As the examples in the first section show, if sociology can sometimes be concerned with regularities (statistical regularities, for instance, as in the classical work of Durkheim on suicide), it can also be concerned by questions taking other forms. When Sombart asks *Why there is no socialism in the United States* at the beginning of the twentieth century, the question bears on a singularity which distinguishes the United States from other countries rather than on a regularity. When Weber investigates the reasons why the Protestant sects are flourishing in the

United States at the end of the nineteenth century, he also raises a question dealing with a singular state of affairs. Tocqueville's work on *L'Ancien Régime* is based on a question dealing with the reasons for the many cultural, political and economic differences which he could observe between Britain and France. Many modern sociological works could be mentioned which, in the same vein, attempt at explaining events, singularities and differences. In other words, explaining regularities is, in the actual practice of sociology, a particular rather than a general objective.

Point 2. 'Methodological individualism', which is a basic principle of the 'Weberian' approach – though the expression itself came after Weber – has been exposed to objections which appear today as of little worth. 'Methodological individualism' (MI) states that, in order to explain a social phenomenon, we have to make it the aggregate consequence of individual actions, and that we have to understand these actions as adaptive responses given by the various categories of actors to their situation. This is exactly what Trevor Roper and Hagen did in the examples summarized in the first section.

A current objection is that the operation of understanding cannot be controlled and is affected by the subjectivity of the observer. But we know from detective stories that even when the actors try to hide their motivations, they can in many circumstances be reconstructed, sometimes with a high degree of likelihood. In the same way, when Tocqueville explains that the absentee landlords were more frequent in France than in Britain, because by leaving their land they could benefit from the fiscal exemptions reserved to city-dwellers, the explanation appears likely.

A second objection currently raised against MI is that it cannot be useful in the study of macrophenomena and that it should be confined to the study of small groups or organizations. However, many examples belonging to macrosociology can be opposed to this objection. Sombart in his study on the failure of socialism in the United States, Trevor Roper in his study of the relationship between Calvinism and the rise of capitalism, Hagen in his study of the Colombian development, Tocqueville in his study of the reasons for the underdevelopment of French agriculture in the eighteenth century all implicitly or explicitly apply the MI principle although they obviously deal with macrophenomena.

A third objection is that MI can be employed only in the case of the societies with an 'individualistic culture', or to use Tönnies' terminology, in *Gesellschaften*. By contrast, it should not be applied in the case of *Gemeinschaften*, i.e. of the societies where the actors are less 'individualized' and should be rather conceived as elements of a whole, of a totality. As a matter of fact, this objection rests on a misunderstanding. That societies and groups can be more or less 'individualistic' is one thing; that

MI should apply to 'individualistic' societies – those where the market mechanisms play a crucial role – is another. Weber for his part considered that what we now call MI is applicable to all societies: in all cases, we have to reconstruct what the individuals have in mind if we want to explain why they manifest certain behaviour. In respect to this discussion, Popkin's work on Vietnamese village communities is particularly useful. The observers of such 'traditional communities' were very often impressed by the fact that the villagers take their collective decisions on the basis of the unanimity rule. Frequently, they drew from this fact the view that the individuals in such communities were, so to speak, dissolved into the group. But a closer look shows that, in such communities, because of the high interdependence between units, everyone can produce 'externalities' unfavourable to the others. Thus, if I change my method of cropping to increase the productivity of my plot, I can reduce the benefits my neighbour draws from his gleaning on my plot. In a system where everybody can impose severe externalities on everyone else, a 'functional' decision rule – i.e. a rule which all would be likely to accept – is the one which gives a veto right to all, and that is the unanimity rule. This example shows that, even in the case of traditional communities, MI can lead to a better understanding of institutions and practices than the nebulous view according to which, in such societies, individuals would not exist as independent units.

Point 3. The examples in the first section show sufficiently that the principle according to which understanding the actions of the actors is an essential moment in every sociological analysis is not incompatible with the Popperian epistemics. Again, this operation of understanding can be rationally controlled. On the other hand, the consequence of a theory built according to the MI principle – again the example of the Weber–Trevor Roper theory – can be confronted to reality along the line of the Popperian epistemics.

Point 4. Several critics, from Halbwachs to Alpert, have shown that the Durkheimian paradigm is not really tenable. Durkheim defines sociological explanation as an operation which basically should attempt at relating facts with facts, observable variables with observable variables. Therefore, Durkheim shows in his *Suicide* that, other things being equal, the rates of suicide are greater among bachelors than among married people, and greater among married people without children than among married people with children. But the correlation makes sense and has a chance to be perceived as being more than an artefact only from the moment when we can interpret it as an effect on the individual feelings, motivations and behaviour of the differences in the three types of situations.

Point 5. In the above four points, I have examined the debate between the Durkheimian and the Weberian paradigms. The four points tend to conclude in favour of the latter. The debate between the 'positivists' (either Weberian or Durkheimian) and the advocates of an interpretive or critical sociology can be alluded to more briefly.

Interpretive sociology starts from the notion of meaning: a work of art has a meaning for us which a piece of wood, for instance, has not. As the sociologist is confronted with meaningful objects, his relationship to the object cannot be the same as that of the physicist, say, to his object. Consequently, sociological methodology should be different from the methodology applied by natural sciences: the operation of disentangling the meaning of a certain institution would be closer in essence to art criticism than to scientific induction.

What is true in this argument is that we can have a relationship with other people which we cannot have with physical entities, i.e. a relationship of understanding in Weber's sense. I can understand that the French landlords in the eighteenth century tried to escape heavy taxes, because I would like to do so if I could. But, as we saw earlier, this circumstance does not contradict the fact that if a sociological theory is to be convincing it has to follow the procedures described by Popperian epistemics. When we say that we have discovered the meaning of an institution, this does not indicate that we have experienced the kind of illumination which we can experience before a work of art. Let us go back to an earlier example, that of the meaning of the unanimity rule in traditional rural communities. Understanding this meaning is not an irrational operation: we understand it once we understand why it is accepted by the actors, given their situation of high interdependence, and once we see the relation between interdependence and the capacity of everyone to impose unfavourable 'externalities' on the others. Or we understand why Colombia developed at the turn of the century once we understand why a group was deterred from using its resources towards activities other than business. Once such an understanding is gained, it can be rationally controlled, as can any scientific hypothesis. Thus, Berger mentions a study where the high birth rates in India were explained by the culturally enforced submission of Indian peasants to traditions. This theory provided an explanation as to why Indian peasants adopt counter-productive behaviour. But the theory was abandoned once more careful field work showed that, in many cases, given the prevailing economic conditions, a family unit was better off with four sons (i.e. within the average eight children) than with, say, one or two, since with four sons two can work on two plots and produce scale economies while two others can work in the next city and bring their salary home.

Thus, the fact that sociology has to do with such notions as meaning and understanding does not imply in any sense that it should proceed in an irrational fashion or use an intuitive methodology, the rules of which would be located beyond any possible discussion and description.

Of course, the words meaning and understanding have another sense in an expression such as the meaning of history, which implies an evaluative view of historical change. So, when they speak of the meaning of institutions or of the meaning of change, the 'critical' sociologists designate an intellectual activity of the normative or of the evaluative rather than of the positive type. This type of activity very often appears in conjunction with that of analysis and explanation. Thus, terms like *alienation* or *anomie* imply a comparison between a given actual social state and a state considered as desirable or ideal. But this does not say that evaluation and analysis cannot be separated from one another, nor that sociological analysis should always be affected by the interests or the value system of the sociologist. Interests and values can orient the selection of the topics or objects he decides to study, but once the object is defined, the analysis can be conducted along the traditional paths of scientific induction.

When 'critical' sociologists want to see in sociology essentially an evaluative activity, or an activity where evaluation and analysis cannot be separated, they take an arbitrary position which simply ignores the impressive body of actual sociological research where value statements do not play any visible role.

3 THE ACTUALITY OF THE WEBERIAN PARADIGM

Since Weber and Durkheim, a great deal of discussion has gone on as to what sociology is about. The discussion has not yet reached a conclusion, as the present fragmentation of sociology shows. Nevertheless, it has progressed, thanks to a number of contributions often classified as belonging to general sociology. In spite of fragmentation, one can speak of a main stream in sociological research, which is, in my opinion, closer to the Weberian approach than to most others.

I will try to document the latter statement by referring briefly to some recent research and research currents. For a long time now the sociology of organizations has given birth to cumulative knowledge. Organizations are analysed as systems of roles, open on an environment with endogenous changes provoked by the aggregation of individual strategies.

The Weberian paradigm has taken a longer time to penetrate in fields of sociology dealing with phenomena located at the macro, rather than at the intermediary, level represented by organizations and small groups. Several classes of macrophenomena, however, have been approached in recent years in the framework of the Weberian paradigm.

Although the following list does not aim at being exhaustive, four classes are worthy of mention:

(a) the study of social movements;
(b) the field of social mobility and stratification;
(c) the sociology of knowledge;
(d) the field of development and modernization.

In the past, the typical sociological approach of social movements was either taxonomic or correlational. In the latter case, it was attempted to relate the apparition of social movements with other variables, such as the economic cycles. Needless to say, the two approaches have increased our knowledge, suggested many ideas, investigations, etc. But in most cases the correlations appeared to be weak, and variable from one context and from one time to another. A third traditional approach is the interpretive approach, the method and scope of which is close to the nineteenth century 'philosophy of history'. In the recent past, however, several works have used the Weberian MI approach, treating social movements as the aggregated outcome of differential strategies. The interest in the approach is, for instance, to account for the fact that the same causes of dissatisfaction or the same collective demands can give rise to different types of social movements and eventually to no social movements. Thus, an often mentioned study explains convincingly why the civil rights movements took a very contrasted form in the north and in the south of the United States.

In the field of social mobility and stratification, again, traditional approaches can be detected. They are either correlational (for example, is mobility greater when economic development is greater or when education is more widely diffused?) or descriptive (how many classes or strata can be detected? which are they? how are they intersubjectively perceived?). And these approaches have given birth to many findings and ideas. Recently, a stream of works have attempted to approach the mobility flux as the aggregate outcome of individual strategies. That approach has brought interesting findings as to, for example, an explanation of the evolution over time of the educational demand or as to the evolution over time of the structure of mobility. But, against a view of the history of the social sciences which would be too evolutionary, it must be mentioned that this approach is present, in *statu nascendi* so to speak, in some earlier works, such as Sorokin's work on *Social Mobility*.

The same Weberian approach has brought interesting results in the field generally called the sociology of knowledge. Traditionally, the sociology of knowledge, as defined by Mannheim attempted to relate the characteristics of intellectual production to broad societal variables. Inspired as it was by the neo-Marxist perspective, it attempted to discover hypothetical relationships between the 'infrastructure' and the 'superstructure' or between class positions and tastes, values, ideas, etc. No doubt the approach has been useful in drawing attention to the fact that social factors can play a role in the production not only of ideologies, but of knowledge. In recent years, this approach has been more and more clearly perceived as a deadlock, however, and several studies can be mentioned which analyse the characteristics of intellectual production as aggregate outcomes of individual answers to situations with given properties. Thanks to these studies, we understand better, for instance, why the notion of *Bildung* took such an eminent place in Germany at the end of the nineteenth century, why Marxism attracted the French at a time when it did not attract the Americans after the Second World War, and so on. The Weberian paradigm also penetrated the field of the sociology of scientific knowledge. We understand better now to what extent scientific production is affected by scientific institutions, i.e. by the individual strategies and objectives developed by the protagonists within the framework of these institutions.

The field of development and modernization, to take a last example, is today very much oriented toward the Weberian paradigm. This is very interesting, since the notions of modernization and development are located at the macrosociological or, to use Parsons' terminology, at the 'societal' level. Traditionally, the correlational, descriptive and taxonomic approaches occupied a large place in this field. Today, one realized progressively that valid statements with a general scope which can be made on development are very rare. Nobody believes any more, for instance, that the theory of the vicious circle of poverty holds generally, since, if it did, Japan and Colombia at the turn of the century should not have developed. Many works on development – probably the most interesting ones – are now oriented towards showing the aggregate, complex and often contradictory effects which result from endogenous or exogenous changes. As in Hagen's study referred to above, these effects are interpreted as the outcome of individual answers to changes in the situation of the actors. Its approach is typical of a stream of research.

Finally, it should be mentioned that a number of works have been produced in the last two decades, developing formal ideas or formal theories where various types of sociological aggregation effects were studied at an abstract level. This line of research, which was suggested by Simmel through his notion of 'formal sociology', had been abandoned for many decades, while it had persisted in economics. Since the very notion of aggregation implies the MI principle, it is not surprising that this line of research was revitalized at a time when the Weberian paradigm came again to the front of the stage.

4 THE SOCIOLOGICAL SUBCULTURES

I have tried above to isolate and identify what I called a mainstream. While this mainstream exists and is developing beyond doubt, it is not powerful enough to create a 'consensus' among contemporary sociologists. In this concluding section, I will present some remarks on the reasons for this situation.

First, sociology is a discipline characterized by the great diversity of its objects: a sociologist can deal with microphenomena and be interested in studying why and how people divorce or are taken into 'asylums' as well as with macrophenomena located at the societal level. In the first case, his methodology can be close to the novelist's methodology, i.e. be of the 'interpretive' kind. In the second, the notion of 'interpretation' has little meaning and can describe only very subjective views as to what 'society' is or should be. But the fact that it has a meaning in the first case is often taken – wrongly – as a proof of the general legitimacy of the 'interpretive' methods.

Secondly, sociologists have various interests in the cognitive sense of the word. Some are rather interested by describing accurately the phenomena they deal with, some in explaining them. Thus a large part of the literature on social stratification is descriptive. The word sociography is sometimes used to characterize this type of 'interest'. In this respect it is useful to note a current linguistic confusion. As certain statistical methods (such as factor analysis or regression analysis) aim at 'explaining' the variance of one or several variables, the sociologists who use them have often the impression that it is sufficient to apply such methods to the data they are concerned with to explain the phenomena corresponding to these data. Actually, these methods have essentially a descriptive function in that they aim to summarize quantities of information into a small set of parameters: basically, applying factor analysis, for instance, to a set of data is an operation of the same logical nature – although it is more complicated – as the operation of summarizing a distribution by its mean and standard deviation. The word 'explain' has obviously a different meaning in statistical usage and in general usage.

Thirdly, sociologists have different representations as to the objectives of sociology. For some of them, it should have essentially a cognitive function and lead to a better understanding of social phenomena. For others, it should have a practical function and contribute directly to social change, for instance by raising the 'level of consciousness' of the public on a certain issue. Very often, those who have the second attitude would develop the argument that sociology – for some reasons – cannot be 'objective' in the sense of other disciplines. But in most cases the argument rests on rationalization rather than on convincing reasons.

Fourthly, the Weberian paradigm has many advantages, as I have tried to show, but it includes costs. As soon as we realize that a social phenomenon should be analysed as the aggregate consequence of actions developing within the framework of a social system with its singularities, we are confronted with the corollary that discovering general empirical statements is a utopian objective, that there are no 'laws' of development, or of social mobilization, and that regularities are not the main thing sociologists should care about. This point was conceived as a sort of evidence by the German classical sociologists, Weber, Simmel and Sombart. But it offends certain deep-seated convictions, according to which sociology, being a nomological science, should be in a position to discover general empirical statements. In the Weberian concept this view is disqualified: sociology can build and use models or formal theories, or formal systems of categories of general scope, but it has to be understood that these mental constructs have to be adjusted to each particular case. Merton's paradigm of functional analysis, Olson's theory of collective action, or Parsons' pattern variables are well-known examples of such ideal constructs. But it is important to see that, as such, they include no empirical consequences. They are rather mental guides which can be used to build particular theories in the proper sense, i.e. adapted to certain empirical cases. Such statements obviously go against current epistemological representation, to be found in Durkheim's work and in many contemporary works as well, according to which a fundamental objective of sociology would be to propose and confirm nomological statements.

Moreover, these mental constructs which are often designated under the label of 'sociological theory' do not constitute a well-integrated corpus comparable, say, to neo-classical economics.

Fifthly, by contrast to other disciplines, sociology is more dependent on national traditions. Durkheim wrote that Simmel's work was 'metaphysical' and the compliment was returned to him. The consequences of the quarrel are still visible.

However – and this is the point I would like to raise in conclusion – a growing consciousness is appearing today of the fact that an agreement can be reached on some points of the older discussions. Thus, the analysis of empirical regularities is seen not as the main task, but as one of the possible tasks of sociologists. On the other hand, it is more clearly perceived that, if methodological individualism is not always applied, the reason is that in some circumstances it can be applied with difficulty, for instance because of a lack of information at the micro-level, rather than its being a theoretical reason, as Durkheim and many others after him believed.

Comment

H. NOWOTNY

I MAINSTREAM SOCIOLOGY AND THE INFLUENCE OF THE FOUNDING FATHERS

Raymond Boudon's analysis of sociology rests in a kind of three-dimensional space: on one side he places what he called neo-Popperian epistemics, on the other, the plurality of methodological paradigms and in the third dimension the Weberian paradigm. The reference to neo-Popperian epistemics simply means that there are certain explanations in sociology which resemble familiar examples from the textbooks of philosophy of science. It is to Boudon's credit that he emphasizes that this type of explanation is not the exclusive domain of quantitative reasoning, but can be qualitative as well, and that neither is it the monopoly of modern or classical sociological research. The first message is therefore clear and simple: sociology can function, as other sciences do. So far, so good.

The second dimension is constituted by the methodological principles which underlie the plurality of paradigms in sociology. Here, Boudon juxtaposes the Durkheimian and the Weberian tradition, and also re-analyses the old opposition between a positivistic and an interpretative approach. While Durkheim was fascinated by the discovery of social regularities that hold independent of the motives, sentiments or behaviour of individual actors, Weber, Simmel and their followers have denounced this 'obsession' as being based upon a futile search for universals. For them, as Boudon puts it, the main objective of sociology is to explain social phenomena as the product and outcome of individual actions. Consequently, understanding (*verstehen*) such motivations becomes one of the prime tasks of sociology. For another group of sociologists, however, the discovery of 'structures', i.e. non-random combinations among social features apparently unrelated to the individual's behaviour or rational action, is considered the most important task. Distinct from, but related to, this great divide is another debate in sociology about the extent to which it can be as objective as other sciences.

It would lead too far to locate these opposing traditions in sociological thinking in the greater context of the flow of ideas in European history. Suffice it to say that the concept of the rational actor, and its subsequent transformation into the principle of methodological individualism, is closely linked to the dominant paradigm in classical and neo-classical economics. The great attraction of the discovery of the 'social facts' is related, especially in France, to the fact that sociology, in order to gain entrance into the French university system in the last part of the nineteenth century, had to demarcate itself clearly from philosophy, which dominated the academic setting. There is yet another tradition, linked to the physicistic leaning of early sociological research, which later was developed more fully within statistics, to seek aggregate regularities. To put it very simply, while these incompatible paradigms differentiated and became dominant paradigms in other disciplines, they lived on in residual form within sociology, incompatible but united under the common label.

Based on these two dimensions, Boudon adds a third that he views as the major line of advance in sociology, and that is the steady progress of the Weberian paradigm embodied in mainstream research which is his expressed preference. For Boudon, the main reason why the Weberian paradigm was successful lies in one of its basic methodological tenets, that of methodological individualism. It states that 'in order to explain a social phenomenon, we have to make it the aggregate consequence of individual actions and that we have to understand these actions as adaptive responses given by the various categories of actors to their situation'. According to Boudon, major advances in sociology can be found by the successive incorporation by this paradigm, of phenomena located at the macro-level, such as the study of social movement, the field of social mobility and stratification, phenomena in the sociology of knowledge and in the field of development and modernization. In all these fields, the successful approach consists in analysing the flux of changes as the aggregate outcome of individual strategies.

I will now voice some of my criticism regarding this systematic exposition of sociology, before moving on to other important points raised by Boudon. In pitching the Weberian against the Durkheimian paradigms (or the tenet of methodological individualism against structuralism) Boudon misses out not only the influence of the paradigm initiated by the third of the founding fathers, Marx, but also one important social development in general and process theories in particular. While it remains debatable whether – following the demise of structural functionalism in the 1960s – one can speak of mainstream sociology being represented solely by the Weberian paradigm, most sociologists would probably agree that the theoretical and methodological influence of the founding fathers, Weber, Durkheim and Marx, is still inordinately pervasive. The ongoing intellectual development of the major themes which were initiated by them and put on the sociological agenda are a proof of both the powerful conceptual framework they developed for the social transformation that occurred in their day and the historical continuation and transformation of major issues identified and analysed by them. What they share to a certain extent, despite many divergencies, is concern with what can perhaps be called a theory of social processes. They remained fascinated by the dynamics of societal transformation and the mechanisms through which this is accomplished. One of their lasting legacies is the task for sociology to diagnose and explain those long-term and unplanned, yet structured and directional trends in the development of social and personality structures that constitute the infrastructure of history in the making. The fact that much of sociological research is focussed today on short-term phenomena and is a-historical in its approach should not make one overlook sociology as essentially concerned with phenomena that have a history and occur in history. Processes of differentiation and integration, of division of functions and of monopolization are related to each other and can be observed in many spheres of social life. They are unplanned and yet proceed in certain directions.

One example for such analysis can be found in the work of Norbert Elias on the process of civilization. In it he shows how the emergence of the modern nation state is itself the result of changing power differentials and configurations between social groups and how, in a lengthening chain of dependencies, control is gradually internalized. The process of civilization is only one of many long-term and 'blind' historical processes which can be analysed as the result of unplanned configurations between different groups that are constantly changing as a result of shifts in the balance of power.

By reconceptualizing the work of the founding fathers and their continuing influence in sociology in process-orientated terms, we reach another dimension hidden in Boudon's analysis, of where to look for advances. The explanation of societal processes remains one of the foremost tasks in sociological theorizing.

2 THE EMERGENCE OF EMPIRICAL SOCIAL RESEARCH

Another important addition has to be made with regard to the context in which the advancement of sociology is taking place. Boudon's exposition is clear, systematic and internally consistent in the sense that, given certain promising methodological principles, we are made to understand why the Weberian paradigm emerges as 'the best' on the road to progress. Yet, his analysis is both a-historical and unrelated to the societal context in which developments of sociological theorizing and research are taking place. The interrelationship between stages in the development of a society and the demand that is generated for a specific type of social science knowledge has led to the rise of empirical social research. Sociology as an academic discipline is somewhat prone to reject this child of its union with public and administrative demands. Institutionally often located outside the established university system, in some cases too close to the commercial sector, empirical social research has developed into a world-wide empire of data production, analysis and monitoring, of policy-advice and evaluation. Short-term oriented in its overall approach, preoccupied with local problems of administrative and policy demands, quantitative, though not exclusively so, in its methods, eclectic in its theoretical premises, anonymous in the sense of being produced by teams of researchers rather than individual authors who seek to impose their intellectual mark on the product, cumulative in the sense of a growing body of 'findings' which make their impact in a more or less indirect yet continuing fashion on policy-making, empirical social research can itself be looked upon as a major advance in the field of sociology. Its empiricism fits in with the logic of action of the dominant administrative–political establishments. Modernization and democratization of administrative and political structures have generated new kinds of social demands that are met by sociology in a distinct institutional, methodological and conceptual framework.

This is not to reanimate the old debate of the autonomy of a scientific field or to determine whether sociology has reached the stage of 'finalization' that, the Starnberg group claimed, would hold for the more mature disciplines. Sociology has enjoyed, in different junctures of its history, varying degrees of autonomy or has been exposed to political–administrative pressure. This obviously depends upon the national policy framework and the readiness with which a discipline, in its cognitive and institutional constituents, responds to them. This is a matter which can be approached through empirical study. However, there appears to be a long-term trend towards the production of more policy-relevant social research which is accompanied by structural changes on the side of the administrative system: the institutionalization of research-orientated units in ministries, departments and other administrative bodies, sector-orientated funding agencies and, on a more general level, contract research. The types of question that emerge within such frameworks and the problems facing policy-makers have made a deep and lasting impact on how sociology in its concepts, method and style of work responds to these demands.

3 SOCIOLOGY BETWEEN THE POLITICAL AND SCIENTIFIC ESTABLISHMENT

In a more theoretical vein one can say that sociology continues to be confronted in an ambiguous way by its relationship to two kinds of establishments: the political–administrative establishment and the hierarchy that exists within the sciences. As a science of social formations, sociology has to be open towards the problems societies face, while it is always in danger of being carried away by ideological winds of change. The old debate on value neutrality and relativism, on objectivity and open partisanship still lives on, as illustrated recently by the discussions about social movements and the researcher's position inside or outside them. This is not only a question of institutional autonomy, but of how the self-conception of a discipline – in this case its relationship to the political powers – is reflected in the kind of knowledge it produces. Since societies have structures and regularities of their own which are relatively independent of human wishes and aims, it is the task of sociology to study them in a sufficiently distanced way. Such an analytical distance is also needed when we observe the anticipatory mechanisms through which the sociological object is conceptualized in different settings.

The other thread which runs through the perennial discussions and disputes so well summarized by Boudon, is related to the anxious question of how 'scientific' sociology really is. This yearning for higher recognition on the hierarchical ladder of scientific reputation takes curious forms: it expresses itself in the imitation of formalized methods or the uncritical use of quantitative data without considering what problems are amenable to such analysis. A curious, unsociological perception exists with regard to what the natural sciences do or what they are: it is as though a *fata morgana* of a timeless scientific ideal would prevail. In sociology's unfortunate attempts to seek the approval seal of academic recognition by imitating the paraphernalia of the higher status disciplines, the sociological imagination is left behind. Sociologists then fail to recognize the extent to which processes of competition shape the cognitive and social structure of a scientific field and the extent to which knowledge production is used as a means to enhance the status and power of a discipline. In their insecurity, sociologists, as the weaker part in a power relationship often does, seek greater security through an act of mimicry, by taking over concepts, methods and forms of thinking of the more powerful. As Michael Pollak has pointed out, in a relationship of domination it is the dominated part that has a greater interest in defining its identity. The dominant part is less concerned about its own, since it usually has more resources at its disposal to manipulate reality in a way that secures the maintenance of its position. The recurrent crisis of identity in sociology is therefore a clear indication of its continuing dependency with regard to the other sciences. In its yearning for being something other than it actually is, sociology expresses its immaturity.

4 THE HETEROGENEITY OF THE SOCIOLOGICAL OBJECT

This brings us back to another tenet underlying Boudon's analysis which is shared by most sociologists, although probably for different reasons, and that is the heterogeneity of its objects of knowledge and, correspondingly, the diversity in analytical perspectives. This is usually considered a weakness, a lack of the consensus which marks more mature disciplines, and evidence for the high vulnerability of sociology with regard to outside pressure or self-imposed deference. I would prefer to turn this view around and look at conceptual diversity as the inherent strength of sociology and, if we manage to understand how and why sociology reconstitutes its objects of studies, even an advancement.

It is not only the diversity of social phenomena or the complexity of the social world which accounts for the heterogeneity in forming collective representations of them. All forms of reality, natural and social, appear extremely complex and diverse if we have the conceptual and methodological tools to study this complexity. Complexity is usually reduced by distance, be it the distance that comes from ignorance or that achieved from knowledge. A peculiar feature of sociologists is that they live amid this complexity and diversity: their views are constantly challenged by others who also live in such diversity and respond to the changing contexts in which it occurs. Whatever analytical perspective is adopted – micro or macro; strategies of rational actors or processes and configurations; the analysis of power and competition as major determinants of social life or the everyday processes of constructing reality and meaning – none of them can claim to grasp exclusively the whole. Rather than bemoan theoretical pluralism, we should enquire under what conditions the object of study is thus divided and how these divisions relate to or oppose each other – in other words, to apply a sociological perspective to what sociologists are doing.

In their well-known article in *Science* of 1971, Deutsch, Platt and Senghaas have identified 62 'leading achievements' generated in the first two-thirds of this century in the social sciences. Their criteria for advance were two-fold: that a new perspective or a new operation had been found and these have proven fruitful in producing a substantial impact that led to further knowledge. Although apparently clear in definition, the application of these criteria produces great difficulties when applied to sociology. I would agree with Boudon that 'methodological individualism' has certain claims in its favour, but the general feeling is probably that it is extremely difficult to point towards a major breakthrough in sociology in a fashion analogous to breakthroughs as they occur in the natural sciences. If one desperately looks for such an occurrence, however, it is ironically not within the discipline itself, but outside: I mean the transdisciplinary impact sociology has enjoyed in the last decade. To an astonishing degree, sociology has shown an intellectual radiance beyond narrow disciplinary *problématiques*.

This is especially true for history; it holds to a certain degree for what is called the social study of science; for art and literature; for women's studies. In all these fields and studies, an ingenious combination of neo-Marxist, Weberian, Durkheimian and other paradigms have wielded a diffuse, but noticeable and lasting influence. Sociology is establishing itself as a cognitively powerful 'residual' discipline from which emanate waves of renewal that throw new light on the social objects and relations figuring prominently in other fields. Its most common denominator is the sociological perspective – one that sees the world essentially as socially constructed; one that recognizes that structures and behaviour have meaning and define the meaning of individual actions; that power and dependence relations are an integral part of any social situation, yet power is never one-sided; that strategies and negotiations are part of the interaction patterns that more often than not are governed by competition and conflict; that processes are often directed, yet unplanned, etc. Such a perspective can be brought to bear in many different spheres of social life and on many different levels. It needs, like any attempt at theorizing, hard empirical work to substantiate it, in order to gain credibility and eventually a synoptic view to render its actual achievements more visible. But the most unique contribution sociology has perhaps made is its introduction of a kind of self-reflexivity which links the cognitive and the social level and makes us recognize that our way of thinking about the world cannot be separated from the objects of study and how we conceptualize them.

5 CONTINUITIES AND DISCONTINUITIES: TOWARDS AN INTERNATIONAL PARADIGMATIC COMMUNITY

Finally, I want to single out the last point mentioned in Boudon's paper – the embedding of sociology into national traditions – as deserving far more visibility and attention. The whole complex of the relationship of sociology to other disciplines, such as philosophy, its history of institutionalization inside and outside the university system, are highly dependent on the national context in which this occurs. The relationships of sociology to the hierarchies of political power are structured by and take place within given national traditions. The way in which sociology is diffused and the kinds of transactions that occur when crossing national boundaries have to be treated in the light of boundaries of culture, languages, political frameworks and the specifics of the institutional settings of the nation states. The national components of scientific work are only now beginning to surface. They pertain to the natural sciences as well, but are more pronounced and paradoxically also more disguised in the social sciences. What could still pass not long ago as universalism in sociology, the paradigm of structural functionalism, appears in retrospect as the hegemonistic claim of one national variant which derived from the then dominant position of American sociology.

These and other questions related to the national and international components in the sciences deserve our common interest and might provide a promising and urgent agenda for a forum such as this. What is needed is the move from national paradigms towards an international paradigmatic community. If it is to be expected that constrained budgets for science will concentrate demand increasingly upon national issues, what will happen to the international side of scientific work? Why are there times – for instance, the contribution of the social sciences in Austria between the wars – when a series of innovative links occur between the concern for national, or even local, problems and yet great openness existed towards international currents of ideas, developments and tendencies? What are the transformation mechanisms that link these two sides, while in other times a relapse into provincialism becomes the seemingly inevitable fate of a particular field in a given country?

There is, as Boudon rightly points out in his conclusions, a growing consciousness of the fact that an agreement can be reached on some points of the older discussions. An interesting attempt at a *rapprochement* has recently been started between Marxism and some of the proponents of modal logic and game theory. Other such examples could be enumerated. There is also growing awareness of the fact that theoretical pluralism is an inherent feature of the discipline which testifies to its vitality. This pluralism derives from sociology's capacity to reconstitute its object of study under different perspectives and to reflect upon the cognitive and social conditions under which this takes place. In a world in which the pressure towards cultural homogeneity is growing at an enormous rate, this is not a vice, but a virtue, provided the process through which it occurs can be made accountable to rational discourse. I see this happening especially at the borderlines between sociology and some of the neighbouring disciplines.

There is also greater awareness of both continuities and discontinuities in the history and construction of the identity of this discipline. Discontinuities arise partly as a consequence of sociology's still precarious, institutionalized existence and its relative vulnerability towards pressures from outside, including its deference towards other establishments. Discontinuities arise because of national traditions of paradigms and the abrupt transformations which occur when sociological thought crosses national frontiers. Continuities follow from the development of some of the main themes put on its intellectual agenda by the founding fathers and which are, when encountered with an open mind, far from obsolete today. Continuity and the accumulation of empirical knowledge are fostered by a growing body of empirical social research. The conceptual anticipation of social demands and of problems posed in administrative terms was translated into sociological problems in a distorted way, but constitutes important empirical background material against which theoretical interpretations can be checked. In its more empirically oriented branch, sociology can even be said to have reached a new kind of stability, due to the continuing demands for societal monitoring. Paradoxically, it is perhaps this institutional stability which provides the necessary security for sociology to continue with its otherwise destabilizing and partly subversive function. By engaging in de-mystifying and critical appraisals of what is taken for granted in a society, sociology remains an open discipline; nothing could be worse than a premature dwindling into a kind of sociologism. Its greatest merit, in my opinion, is that it has preserved something of the passion for the hypothetico-possible.

Although passion may not be the most fashionable topic these days, let me remind you that it too can be analysed from a sociological perspective. It then becomes, in the words of one of the foremost German theoreticians, one of the mechanisms through which society advances in its evolutionary development by 'increasing the likelihood of that which is unlikely to occur.'

Progress in linguistics

S. C. DIK

> Virtually every major period in the history of science is characterized both by the coexistence of numerous competing paradigms, with none exerting hegemony over the field, and by the persistent and continuous manner in which the fundamental assumptions of every paradigm are debated within the scientific community. *(Laudan, 1977)*

I SOME PROBLEMS FOR LINGUISTICS

LINGUISTICS is concerned with the study of natural languages in their various manifestations. Natural languages have been studied for centuries, from many different angles and with many different purposes. Nevertheless, we do not seem to be anywhere near a full understanding of this all-important phenomenon. This very fact poses some intriguing problems to linguists. Let me mention some of these enigmas:

(a) Every normal child acquires an adequate competence in his native language within a few years, without apparent effort. On the other hand, no combined effort of linguists of whatever persuasion has so far been able to unravel all the rules, principles and strategies which together constitute that competence. Is even the most clever professional linguist outwitted by any child of 10?

(b) Every normal adult is an expert user of natural language. We can all use language in remarkably efficient ways. Reflecting on what principles underlie this practical ability, however, appears to be quite a different matter. What people, including linguists, say about language may be a far cry from what they actually do. What makes it so difficult to set out the underlying machinery of that practical linguistic ability in which every normal adult can be rightly called an expert?

(c) Several thousands of natural languages are spoken in the world today. These languages exhibit quite a bit of variety in structure. On the other hand, there is good evidence that they share a common ground plan, that they are all built according to a common design. What are the basic properties that a communicative system must possess in order to qualify as a natural language? Even though we seem to be getting closer to an answer to this question, lots of hard empirical and conceptual work will have to be done before the answer could be satisfactorily finalized. And then, of course, a higher-level question remains: why is it that all natural languages share exactly these properties among all the properties which human communicative systems might conceivably have?

(d) All natural languages change continuously, although speakers are usually unaware of this, and can certainly not be said to contribute consciously to such change. A single language may develop into several different varieties which, after some time, may even be mutually unintelligible. It will thus be very difficult to predict in which direction a given language will develop. Nevertheless, we can identify certain possible changes, which together define limits within which language change can take place. But why do languages change? And which factors contribute to one type of change rather than another? Again these are questions to which we can give only tentative answers. We can describe language change, but we do not fully understand it.

These are only some of the questions which we can express but are unable to solve with the present state of our knowledge. Why is it important to study such questions?

2 THE IMPORTANCE OF LINGUISTIC RESEARCH

Linguistic research is important for both theoretical and practical reasons. On the theoretical side, we wish to arrive at an understanding of how natural language is structured and of how it works, through reconstructing the network of rules and principles which underlie its structure, functioning and historical development. It is sometimes suggested that such a purely theoretical aim is no justification for spending large sums of research money. But that is certainly a limited, and even dangerous view. Human beings have rightly been characterized as 'speaking animals': natural language to a large extent defines the essence of human nature. Human psychological and social life, culture and history, and thus even science itself essentially depend on the possession of the faculty of language. The statement that insight into the nature of human language is not a legitimate goal in itself can only be maintained by those who believe that there is no reason to try to understand human nature. I do not believe that such a view needs any refutation in the present context.

A second reason why it is important to gain insight into the workings of human language lies in the many potential applications of such insight to practical problems involving language. Scores of people have to tackle such problems every day: in learning and teaching languages; in translating from one language into another; in trying to improve linguistic communication; in diagnosing and remedying language disturbances; in planning and implementing language policies. It is a reasonable assumption that all such practical problems can be tackled more effectively if they are approached with a better understanding of the fundamental properties of natural language. Most probably, they can only be solved satisfactorily on the basis of such an understanding.

3 ASSESSING PROGRESS: SOME PROBLEMS

It is not easy to discuss the assessment of progress in a complex field of enquiry such as linguistics. I would therefore like to start by outlining some of the problems involved in this task.

First, evaluation of a certain scientific development in terms of progress, stagnation or even retrogression is a matter of judgement. As with any type of judgement, our evaluation will depend on our positions with respect to a number of basic issues, assumptions which define the 'philosophy' in which we approach the scope and aims of scientific research in our field of enquiry. These basic issues include the following questions:

(a) How wide or narrow is the definition of the subject-matter of the field in question?
(b) How does one conceptualize the nature of the object of enquiry?
(c) How does one define the aims and goals of scientific research in the given field?
(d) By what means does one believe that progress can be achieved with respect to these aims and goals?

Secondly, if the assessment of progress is a matter of judgement, and if such judgement is dependent on an underlying philosophy, then the question arises whether anyone is able to draw an objective picture of the type and rate of advancement in a given field. I, for one, believe that no such objectivity can be achieved.

By this I do not mean, of course, that the evaluation of scientific endeavour should be left to the mercy of private whims and idiosyncracies. What I mean is that no one can disentangle himself from the system of values which, consciously or unconsciously, enters into any conception of scientific work. Even the philosopher of science, whose task is to describe and explain how sciences develop and fail or succeed in achieving their aims, is bound to do so in terms of conceptions of his own. It makes a difference whether he takes a normative or a purely descriptive view of his task; whether he believes that scientific development should be evaluated in terms of some absolute standard of truth, or in terms of some measure of intersubjective agreement; and whether or not he believes that the development of scientific work is dependent on psychological, sociological and even political factors.

If even those who take a meta-position with respect to actual scientific work are bound to have their vision coloured by paradigmatic conceptions (if, in other words, Thomas Kuhn's notion of 'paradigm' illustrates itself), then *a fortiori* a participant in the actual development of a given field should be aware of the amount of subjectivity which enters into his judgements. Instead of posing as an objective judge, he should take measures to avoid being interpreted as such. Such measures may include the following:

(a) he should state in which respect personal conceptions enter his judgements;
(b) he should relate his own views to these personal conceptions;
(c) he should indicate to what extent such personal conceptions are ratified by interpersonal agreement in this field.

On the basis of these considerations it will be clear, I think, that I intend to give a personal view of the achievement of progress in linguistics; at the same time, by paying heed to the three principles above, I hope to avoid the danger of idiosyncracy.

4 SOME BASIC PARAMETERS

What one considers to be progress in a given field is dependent on the positions one takes with respect to the questions on page 117. I will now consider these questions.

4.1 The scope of the field

(a) How wide or narrow is the definition of the subject-matter of the field in question?

In general, the importance of this question seems obvious: any scientific activity requires some delimitation of its problems. Of course, the delimitation may change as research proceeds. But it is clear that a narrow delimitation of the problem may achieve results which, on a wider delimitation, might be evaluated quite differently: local progress in a limited field may contribute little to progress in other areas.

In linguistics, this question is particularly important. Natural language is an intricate phenomenon, with many different manifestations. It can be conceived as a system of rules, which define a set of linguistic expressions; but it can also be conceptualized as an instrument which is used according to specific conventions and strategies. It can be approached as a synchronically functioning system of conventions; but it can also be studied in its development through time. It can be treated as an autonomous system; but one may also incorporate its psychological manifestations. It can be regarded as a property of the individual human being; but it can also be viewed as a social property of a linguistic community as a whole.

Thus, linguistics in the wide sense of the term covers a vast area of potential research, which can only be properly worked with the help of quite a number of disciplines. It is no wonder, then, that some of the most influential theorists of language have felt compelled to cut down the discipline to manageable size by restricting the domain of 'real' linguistics to what they considered to be the essential core of natural language.

This strategy was followed by Ferdinand de Saussure in his epoch-making *Cours de linguistique générale* (1916), where he separated the linguistic system (*langue*) from the use of language (*parole*) within the overall linguistic phenomenon (*langage*), and declared *langue* to be the primary object of linguistic theory, in terms of which all other manifestations of *langage* were to be evaluated. This move was inspired not only by a wish to define a homogeneous object for linguistic theory, but also by a desire to claim an autonomous territory for linguistic research, a territory linguistics would not have to share with other disciplines:

> . . . if we study language simultaneously from different angles, the object of linguistics looks like a confused conglomerate of heterogeneous things without any connecting links between them. Proceeding in that way one opens the door to various disciplines – psychology, anthropology, normative grammar, philology, etc. – which we wish strictly to separate from linguistics but which, due to this incorrect method, could reclaim language as one of their own objects of inquiry.

The same strategy was followed by Noam Chomsky, when he restricted linguistic theory essentially to the theory of grammar (corresponding to Saussure's *langue*), for which he claimed psychological relevance in the form of grammatical *competence* (tacit knowledge of grammatical rules), as distinguished from *performance*, the actual use of language in concrete situations:

> A grammar of a language purports to be a description of the ideal speaker-hearer's intrinsic competence. (1965)
> Linguistic theory is concerned primarily with an ideal speaker-listener, in a completely homogeneous speech-community, who knows its language perfectly . . . (*Ibid*)
> It seems natural to suppose that the study of actual linguistic performance can be seriously pursued only to the extent that we have a good understanding of the generative grammars that are acquired by the learner and put to use by the speaker or hearer. (1964)

It is evident that progress in the area of *langue* or *grammar/competence*, so delineated, must in the first instance be determined in terms of the self-imposed limitations of these theorists. The fact that such progress may have no bearing on the further clarification of psycholinguistic or sociolinguistic problems is, for them, irrelevant, since they have on purpose excluded such problems from the area which they have assigned to linguistic theory.

However, the restriction of linguistic theory to an abstract system of grammar may turn into a strait-jacket for those who would wish linguistics to have an explanatory potential with respect to actual human linguistic behaviour. A straightforward theory of grammar does not tell us how speakers and listeners actually use their language; it sheds no light on differences in competence which may have far-reaching social consequences; it gives us no information on how language-users can achieve fruitful communicative interaction through language; and it has little to say about language as an integrated component of cultural patterns. The progress achieved within the limits of the theory of grammar may thus be evaluated as stagnation by those who wish linguistic theory to respond to questions of a psychological, sociological or anthropological nature.

My personal view is that linguistics, or linguistic theory, should not be so narrowly delimited as is the case in Saussure's and Chomsky's view. For one thing, in such a narrow view, who is going to take care of the many aspects of language which fall outside the limits of grammar? Who is served by the simple statement that the use of language, its social and regional variability, its interpersonal and cultural functioning do not belong to the field of linguistics 'proper'? More importantly, the question may be raised whether a full understanding of the essence of natural language is possible if one isolates the abstract system from its actual functioning in real-life situations. But this is dependent on the answers given to the next question.

4.2 The nature of the object

(b) How does one conceptualize the nature of the object of enquiry? *In general*, the relevance of this question for the evaluation of progress in a discipline again seems obvious.

In linguistics, the relevance of this question emerges quite clearly when one considers the various preliminary definitions of 'language' which different theorists provide.

We saw that Saussure, within the heterogeneous set of phenomena of *langage*, immediately delimited *langue* as the proper domain of linguistic research. This is how he defined *langue*: 'it is a system of signs in which nothing is essential other than the union of meaning and acoustic form, where both these parts of the sign have psychological status.' In Saussure's further elaboration of *langue* as a system of signs, the main idea is that the value of each sign is determined by its relation to other signs in the system: it is what it is by virtue of being different from other signs.

This view on the fundamental nature of the linguistic system defined the programme for a *structuralist* approach to language, which had great impact inside as well as outside linguistics; it defined the programme for a *semiotic* approach to language, since it conceptualized language as a specific instance of the more general notion of a system of signs; and it defined a *mentalist* programme for linguistics, insofar as the central concept of the sign is given a psychological interpretation.

Now compare the initial definition of 'language' as given by Edward Sapir (1921), who wrote in the anthropologically oriented research tradition initiated in the United States by Franz Boas that 'language is a purely human and non-instinctive method of communicating ideas, emotions, and desires by means of a system of voluntarily produced symbols.' Note that language is not defined here as a system of symbols, but as a method of communicating by means of such a system. This gives us a dynamic rather than a static concept of language, a goal-oriented rather than an autonomous view of the linguistic system, which naturally leads to a functional rather than a formal conceptualization of language. Furthermore, Sapir takes a clear stand in the age-old debate as to whether human language is a natural or a cultural phenomenon: in his view, speech is a non-instinctive, acquired, cultural human function, built up through unconscious creative effort of a long series of generations within a given community. It is no wonder that, through Sapir's conceptualization of language, interest is turned to the relations between language, personality and society and that, in contrast to Saussure, it was Sapir's stated aim to cultivate the interdisciplinary study of language.

Sapir's important contribution was somewhat overshadowed by the work of Leonard Bloomfield, who derived his inspiration from mechanist philosophy and behaviourist psychology. This is how Bloomfield (1926) defined language: 'An act of speech is an *utterance* . . . The totality of utterances that can be made in a speech-community is the *language* of that speech-community.'

Note, first, we have here an extensional definition of 'language': language is not defined as a system (Saussure) or a method (Sapir), but as a set of utterances. Secondly, each utterance is a concrete act of speech. The act of speech, in turn, is characterized in terms of observable vocal features on the one hand, and stimulus-reaction features on the other. The former constitute the form, the latter the meaning of utterances. In this behaviourist programme, meaning is thus defined in terms of observable behavioural properties. This had important consequences for Bloomfield's view of the relationship between form and meaning. Since meaning can consist of any feature of human behaviour and the outside world, the study of meaning cannot be the task of linguistics: it is the task of all sciences taken together. Thus, although meaning is essential for understanding language, linguists can do no more than take meaning for granted, and concentrate on the analysis of form. The task of linguistics is therefore to identify and classify the forms of language, and to study the distribution of these forms with respect to each other. Needless to say, this behaviourist programme leaves no room for Sapir's 'ideas, emotions, and desires', nor for the Saussurean interpretation of the form and meaning of signs as psychological phenomena.

For a last example of how language can be conceptualized, consider the following definitions from Noam Chomsky's *Syntactic structures* (1957), the book which was to have such a revolutionary impact on linguistic theorizing:

From now on I will consider a *language* to be a set (finite or infinite) of sentences, each finite in length and constructed out of a finite number of elements . . .
The grammar of L will thus be a device that generates all of the grammatical sequences of L and none of the ungrammatical ones.

At first glance, Chomsky's conceptualization of language is quite close to that of Bloomfield. Chomsky also gives an extensional characterization of 'language'. However, there are important differences: the elements of language are not utterances (concrete space–time occurrences), but sentences (abstract constructions underlying such occurrences). This signals Chomsky's return to a mentalist concept of language: utterances are things that can be observed; sentences are objects that can be known, but not observed. A second essential difference is that, in Chomsky's conception, attention shifts from language to grammar, interpreted as a generative system of rules capable of producing the language. This idea of viewing grammar as a production system in the mathematical sense of the term would prove to be enormously productive in the subsequent development of linguistic theory.

With Chomsky's mentalist interpretation of grammar and his emphasis on abstract underlying principles, one might have expected him to revert to a position in which form and meaning are equally essential determinants of linguistic organization. This, however, he did not do: he continued post-Bloomfieldian formalism in his statement (1957) that 'grammar is autonomous and independent of meaning'. In fact, the autonomy of syntax is one of the central tenets of Chomskyan linguistics.

I have briefly shown four different conceptualizations of 'language':

(a) language as a system of signs, held together by their mutual relations (Saussure);
(b) language as a method of communicating by means of a system of symbols (Sapir);
(c) language as a set of utterances, each consisting of observable vocal and stimulus-response features (Bloomfield);
(d) language as a set of abstract sentences, for which the grammar is to reconstruct the underlying organization (Chomsky).

It will be evident that such different conceptualizations of the object of enquiry will necessarily define quite different research programmes for linguistic theory, and will thus also lead to different evaluations of what constitutes progress in linguistics. The impact of such differences can be amply witnessed in the history of linguistics of this century.

4.3 The aims and goals of research

(c) How does one define the aims and goals of research in the given field?

In general, it would seem natural to define progress in terms of closer approximation to the stated aims and goals of the enquiry. Different formulations of the aims and goals may thus lead to different evaluations of degrees and types of progress.

In linguistics, it is not difficult to come up with a formulation such as 'the aim of linguistic research is to describe and explain the phenomena of natural language'. Such a formulation, however, does not carry us very far. In section 4.1 we saw that there are quite different ways in which the problem area of linguistics can be delineated, and in 4.2 we saw the object of enquiry can be conceptualized in different ways. Such differences, of course, also imply varying formulations of the aims and goals of linguistic research for fields within the total domain, and for philosophically distinct approaches to the subject-matter within each chosen area. From one point of view, one will concentrate on the structural properties of language, as distinct from their functional implementation in wider contexts of use. From another point of view, one may concentrate on properties of natural language which can be studied only within wider contexts of use. What we see in current linguistic theory is a proliferation of subfields and paradigms, each with its own objectives. Consider the following examples:

In *generative grammar* one wishes to explain the underlying cognitive organization of human beings which enables them to form an adequate grammatical competence on the basis of quite limited input data. It is conceptualized in the form of Universal Grammar, a system of innate linguistic principles of quite general bearing, common to all human beings, which together define the essential ground plan for natural languages.

In *functional grammar* one wishes to understand how communicative verbal interaction is made possible by the possession of linguistic knowledge, and, conversely, to what extent the organization of natural language is determined by conditions imposed on its communicative use.

In *historical linguistics* one wishes to understand the basic principles according to which languages develop through time, what internal and external factors determine such development, and how in the course of time languages may split up into different languages.

In *psycholinguistics* one wishes to understand the psychological properties and mechanisms which underlie the human ability to produce, comprehend, store and retrieve linguistic information.

In *sociolinguistics* one wishes to comprehend how socially determined variations in linguistic behaviour, and differences in attitude to such variations, determine the social position and the social mobility of speakers on the one hand, and the historical development of language on the other.

In *speech act theory* one wishes to understand how it is that 'people do things with words', that is which communicative acts can be performed by means of linguistic expressions of given properties, and what conditions underlie the success or failure of such acts.

In *conversation analysis* one takes actual, 'live' everyday conversation as the phenomenon to be understood, where conversation is conceptualized as an activity pattern in which human beings create their social identity with respect to each other, and in which they find the means of dealing with the problems of everyday life.

This list could easily be extended by further subdisciplines and approaches in linguistics, each one with its own aims and goals, each requiring its own methodology, and each with its own definition of the phenomena to be described and explained.

It seems clear, then, that there is no single aim common to all those who are concerned with the study of natural language. Language is a many-sided phenomenon, and a number of distinct research goals may be rationally pursued. Contributions to the study of language should first of all be evaluated in terms of their own stated aims and purposes. At a higher level of evaluation, the question arises as to what contribution each subfield, and each approach within a subfield, has to offer to the overall aim of arriving at an understanding of natural language in all its manifestations. My personal view is that it would be premature and non-productive to claim the sole right to the title of 'linguistic theory' for any current subdiscipline or approach. Rather, the various strands of linguistic research should strive towards arriving at an integrated view of the many facets of natural language.

4.4 How does scientific research progress?

(d) By what means does one believe that progress can be achieved with respect to these aims and goals?

In the preceding sections I have sketched a pluralistic picture of linguistic research. Such research may take quite different forms, according to the problem area which one singles out for special attention, the manner in which one conceptualizes the object of enquiry, and the aims and goals one defines for linguistic research. These parameters which, as we have seen, interact in part with each other, together define a number of distinct subdisciplines within the overall field of linguistic enquiry. Some of these subdisciplines complement each other (to the extent that they address different aspects of the total linguistic phenomenon); some of them compete with each other (to the extent that they define different perspectives on the same aspect of language).

How could one speak of progress in such a fragmented domain of research? One would be inclined to turn to the philosophy of science for an answer to this question. And, indeed, many a linguist has tried to find the justification for his particular view of linguistic research in ideas emanating from that meta-discipline. In particular Kuhn's (1962) contribution to the analysis of scientific revolutions has had its influence on discussions about progress in linguistic research.

Briefly summarized, Kuhn's view boils down to the following theses:

For a discipline or field of interest to develop into a 'science' it is essential that the general acceptance of a common 'paradigm' in terms of which the problems to be solved in that field can be articulated.

A paradigm consists of a set of basic assumptions concerning the goals and purposes of the discipline, and a 'model' of how, given such basic assumptions, relevant results can be obtained.

'Normal science' consists of problem-solving activities within the limits of a given paradigm, the correctness of which is taken for granted.

A scientific revolution is a switch from one paradigm to another. This can occur only when either the reigning paradigm has met with a sufficient number of 'recalcitrant' problems, which it either cannot solve, or can only solve in an *ad hoc* fashion, or an alternative paradigm has been outlined, which seems capable of solving such problem cases, while at the same time handling those in which the earlier paradigm was successful.

Should this view be correct as a general account of scientific progress, then we would have to conclude that linguistics has not yet achieved the status of a true science: it does not display the general paradigmatic unity which, in Kuhn's view, is a precondition for 'normal science' to develop. Some linguists seem to accept this implication, suggesting that linguistics is still in the pre-paradigmatic natural history stage of data-collecting and ideological quibbling. The next logical step would then be the recommendation that everyone should get off his private hobby-horse in order loyally to embrace the most promising research paradigm; not surprisingly, it would coincide with the personal view of the recommender.

A second line of reasoning might be that Kuhn's theory, correct as it may be for developments in the natural sciences, which constitute its main problem domain, is not in the same way applicable to the linguistic sciences. This opinion has been voiced by such authors as Hymes (1974) and Percival (1976). Both agree that Kuhn's view sheds an interesting light on some aspects of the development of linguistics, but that it breaks down on the fact that linguistics, in no phase of its development, has known a period of complete paradigmatic unanimity. Rather, the history of linguistics can be pictured as consisting of a series of 'waves' which, for longer or shorter periods, may have commanded the greatest amount of attention without, however, a single wave prevailing over all the others. Linguistic paradigms seem to die hard and to revive at unexpected moments!

Even more interesting, of course, is the possibility that Kuhn's mono-paradigmatic theory of 'normal science' does not even provide a correct account of progress in the natural sciences; in this respect we can learn from subsequent discussions on the philosophy of science itself. Indeed, critics of Kuhn's position have been quick to point out three shortcomings of his concept of paradigmatic unanimity as a criterion for 'normal science':

(a) it is historically incorrect even for the natural sciences (see the quotation from Laudan, 1977, which I took as a motto for this essay);
(b) it would, if correct, logically preclude any progressive switch from one paradigm to another;
(c) it does not even constitute an ideal state that every science should strive to attain (at least not for those who believe that critical thinking is an essential ingredient of any progressive scientific activity, cf. Popper, 1970, and other contributions to Lakatos and Musgrave, 1970).

These critical points are treated at length in Laudan (1977), whose alternative view on scientific progress has been a source of inspiration for the present paper. The point is that Laudan's theory of progress, which is again primarily, though not exclusively, meant to cater for the natural sciences, is applicable almost *in toto* to the history of linguistics as I see it. Let me mention the main elements of this theory:

Any science is, at every moment of its existence, characterized by several different research traditions.

'A research tradition is a set of general assumptions about the entities and processes in the domain of study, and about the appropriate methods to be used for investigating the problems and constructing the theories in that domain.' (1977).

'A successful research tradition is one which leads, via its component theories, to the adequate solution of an increasing range of empirical and conceptual problems.' (*Ibid*).

'The choice of one tradition over its rivals is a progressive (and thus a rational) choice . . . to the extent that the chosen tradition is a better problem solver than its rivals.' (*Ibid*).

'The evaluation of theories is a comparative matter. What is crucial in any cognitive assessment of a theory is how it fares with respect to its competitors.' (*Ibid*).

Note that, whereas in Kuhn's view the existence of alternative paradigms is a sign of scientific immaturity, in Laudan's perspective it is a necessary condition for scientific advancement. This not only saves us from the self-denying conclusion that linguistics is not really a science, but it also points to paradigmatic disagreement as a means of, rather than an impediment to, scientific progress.

Science certainly requires co-operative activity, carried out with great tenacity by researchers operating within the same research tradition. On the other hand, it also needs the creative proliferation of alternative approaches, if only to prevent researchers from dozing off in complacent unanimity (see Feyerabend, 1970).

5 A COMPONENTIAL ANALYSIS OF RESEARCH TRADITIONS IN LINGUISTICS

The development of linguistics may be compared to a rope, composed of a number of distinct strands, each strand representing a certain research tradition. At each moment in time, some strands may be stronger than others, and thus contribute more substantially to the strength of the rope. Some strands may wither away in the course of time, and new strands may be twisted into the rope, sometimes from outside linguistics proper.

Research traditions are not necessarily incompatible and thus in competition with each other. They may be complementary in the sense that they address different aspects of the linguistic problem area. Complementary research traditions can peacefully coexist without hindering each other; competitive research traditions may, and do, also coexist, but usually in a less peaceful manner: they represent approaches which, in the long run, cannot be assumed to be correct at the same time. To take research traditions as unanalysable monolithic wholes, however, does not do justice to some important properties of progress in linguistics: first, different research traditions may be alike in certain respects, and secondly, research traditions normally do not disappear without leaving an impact on subsequent developments.

We can say that the different strands which form the linguistic rope themselves consist of component threads; that given strands may share certain threads; that certain threads may be twisted into new strands, or may disappear from given strands; and that new combinations of given threads may produce new strands.

What are the components of which research traditions may be said to consist? I shall take these components to be positions which the linguist may assume with respect to a number of basic questions concerning the scope, aims and methods of linguistics. In order to arrive at a componential analysis of linguistic research traditions, it is necessary to go into these questions in more detail. In doing so, I will borrow some concepts from componential analysis as it is applied in linguistics on the level of sound (in the analysis of phonemes into distinctive features) and meaning (in the analysis of meanings into semantic components). In both cases, the units to be analysed are taken to differ from each other along a number of dimensions, on each of which they can take different values. For instance, phonemes can differ from each other along the dimension by VOICE by taking the values *voiced* (+ Voice) or *voiceless* (− Voice); and meanings may differ along the dimension of ANIMACY by taking the values *animate* (+ Animate) or *inanimate* (− Animate).

In a similar way, research traditions may differ along a number of crucial dimensions on which they may take different positions. Sometimes these positions are contradictory; sometimes they are contrary in the sense that they highlight opposite ends of a spectrum, without necessarily excluding each other. In the latter case, the contrary positions may allow a higher-level synthesis in which both of them receive due attention.

In discussing the different component principles of linguistic research traditions, I will also try to indicate to what extent these principles have been productive in contributing to advancement.

DIMENSION I: the system and its use

> Position (I:1). The system of a language is autonomous with respect to the ways in which it is used; in studying the system, one should abstract from conditions of use; in order to understand the conditions of use, one must have an antecedent analysis of the system.

We saw in section 4.1 above that this position was explicitly advocated by both Saussure and Chomsky. In practice, if not in theory, it was also adopted in Bloomfieldian linguistics. The autonomy thesis has thus been enormously influential in shaping linguistics into an independent field of its own, and this was, in part, the very reason why it was adopted.

> Position (I:2). The system of a language is an instrument for communicative interaction; it is thus a system to be used for specific purposes; a full understanding of the system cannot be arrived at if it is cut loose from its conditions of use.

This functional or pragmatic interpretation of the linguistic system is at least as old as the autonomous interpretation embodied in Position (I:1). We find it in Sapir's approach to language as sketched in section 4.2 above, and more generally in the anthropologically oriented research tradition which he inspired. Though perhaps less popular than the Bloomfieldian and later the Chomskyan traditions, the pragmatic position has been persistently present in American linguistics through the work of such linguists as Kenneth L. Pike and Dell Hymes, leading to increasing interest in the pragmatic determinants of linguistic organization in the last decade. In European linguistics, the Prague School contributed to a functional reorientation of Saussurean structuralism in order to arrive, in the words of Roman Jakobson, at a 'means-end model' of linguistic organization. This functional orientation was further developed in France by André Martinet. The British contributions of J. R. Firth and M. A. K. Halliday can also be interpreted in terms of the pragmatic position.

The study of language as a means for communicative interaction was further strengthened by various contributions from outside the field of linguistics proper. The British philosopher J. L. Austin, building on the tradition of ordinary language philosophy, turned his attention to the analysis of what communicative acts speakers may perform in using language, and to the conditions which underlie the success or failure of such acts. This was turned into what was later termed *speech act theory* by the American philosopher J. R. Searle. This work had great influence on those linguists who espoused the position (I:2), and helped to turn the pragmatic approach to natural language into a powerful research tradition of its own.

Another contribution came from the so-called ethnomethodological tradition in sociology. In this tradition, attention focusses on the methods and strategies by means of which people shape their mutual interaction in everyday life. Since verbal communication is the most important ingredient of such interaction, the analysis of normal everyday conversation became an important topic on its own from this point of view.

In different respects, the pragmatic position thus seems to be gaining in momentum. This is not to deny the salutary effect which concentration on the linguistic system as such, as advocated through position (I:1), has had on the search for the principles underlying the system of natural language. Few linguists today, however, would subscribe to the thesis that the study of the ways in which natural languages can be used in communicative interaction should not be reckoned as within the domain of linguistics 'proper'.

DIMENSION II: the relation between form and meaning

> Position (II:1) (the semiotic position). The essential character of the linguistic sign is that it is composed of a form and a meaning; neither can be profitably studied without taking account of the other.

This, as we saw in section 4.2 above, is the Saussurean position with respect to this dimension. It is a persistent feature of all brands of European structuralism, which has never embraced the a-semantic formalist position characterizing some American research traditions. This is one of the reasons why it is fundamentally incorrect to use the term 'structuralism' without distinction for all pre-Chomskyan traditions. Either this term should be used as a general (and thus pleonastic) cover term for all modern linguistics (as in Hymes and Fought, 1975); or, if the term is used in contrast to the Chomskyan tradition, one should take care to avoid glossing over the fundamental differences between the various pre-Chomskyan strands.

> Position (II:2) (the behaviourist position). The forms of language essentially depend on their signalling difference or sameness of meaning; meaning, however, lies in concrete behavioural responses and in features of the outside world; thus, meaning as such cannot be studied by linguistics.

This Bloomfieldian position may have had some temporary usefulness in forcing linguists to attend to the intricate patterning of linguistic forms as such, without paying heed to their meanings. In the long run, however, it has been just as counterproductive as the more general behaviourist position from which it derived. Linguistic research which excommunicates semantics easily develops into empty formalism. This position certainly has not contributed to our gaining insight into the all-important semantic side of language.

> Position (II:3) (the formalist position). The basic principles underlying the formal (in particular, the syntactic) organization of natural languages are autonomous with respect to semantic distinctions; thus, semantic analysis can only be developed on the basis of an antecedently given, independently developed, formal analysis.

This position, claiming an autonomous status for formal syntax, is a basic article of faith of Chomskyan transformational theory. In part, it continues the a-semantic orientation inspired by the behaviourist position (II:2). But this cannot be the whole story, since Chomsky vehemently and effectively combated behaviourism on all other counts. Another powerful influence has certainly been the formalist trend in mathematical logic. Illustrative of this influence is the fact that much of Chomsky's earlier position can be retraced to Rudolf Carnap's *The logical syntax of language* (1936).

There can be no doubt that Chomsky's development of transformational syntax has been the single most influential force in linguistics of the last 25 years. It produced an enormous progress in scope, in that the field of syntax, which had hardly been touched in earlier research traditions, was now disclosed in its remarkable complexity. And it produced progress in depth through its concentration on grammar as a system of interacting rules, guided by deep underlying principles. It is a legitimate question, however, whether this progress was achieved thanks to, or in spite of, the formalist assumption embodied in position (II:3) above. At any rate, relations between form and meaning continue to be a disputed topic even for those who, in principle, endorse this formalist assumption; and they have led to the formulation of several alternatives to transformational theory by those who would be inclined to reject that assumption.

I suspect that many a linguist's dearest wish would be to be capable of integrating the formal sophistication of Chomskyan syntax into a model in which semantics would be handled with equal sophistication. This may in part explain the attraction of the research line initiated by the logician Richard Montague (a tradition now known as Montague grammar); whose aim it is to adapt the means of formal logic to the analysis of natural language in such a way that, for each syntactic operation, a corresponding semantic rule is formalized, which translates the syntactic structure into a logical language which, in turn, can be subjected to the usual principles of semantic interpretation in the logical sense of model theory.

In general, I believe that the implications of Saussure's semiotic position (II:1) have been insufficiently integrated into some of the most influential research traditions of recent linguistics.

DIMENSION III: synchrony and diachrony

> Position (III:1). The synchronic analysis of language must be sharply distinguished from the analysis of its historical development; the latter can only be profitably undertaken in terms of an antecedent theory of the former.

This Saussurean position laid understandable emphasis on synchrony in its attempt to counterbalance the excessive attention to the historical development of languages of nineteenth-century linguistics.

> Position (III:2). The synchronic and the diachronic study of natural language must be carried out in close mutual dependence: a synchronic phase of a language is necessarily a transitional segment of an ongoing diachronic development; the diachronic development is an evolution of the system from one synchronic phase to another.

This synthetic view of synchrony and diachrony was advocated by Prague School theorists. It has gained power in more recent times mainly in two research traditions: in sociolinguistics, which regards synchronic variations in linguistic habits as indications of change in progress; and in typological linguistics, which studies the different types of languages, not only for their own sake, but also with respect to the question of how a language of a given type may change into a language of another type. Each of these developments has given new impulses to the study of language change, and contributed to the more integrated view on the relation between synchrony and diachrony as expressed in position (III:2) above.

DIMENSION IV: linguistic variation

Linguistic habits vary among individuals, socially and regionally defined groups, and styles of speaking. What is the relevance of that variation for linguistic theory?

> Position (IV:1). Linguistic theory can act as if such variations do not exist, and abstract to an ideal, completely homogeneous speech community.

On this Chomskyan view (see section 4.1 above) there is no place for linguistic variation as an independent element in the theory of grammar.

> Position (IV:2). Linguistic variation plays an important role, both in defining the social positions and the social mobility of speakers, and in contributing to linguistic change.

Of course, linguistic variation had been studied for centuries, but mainly in dialectology, which strongly concentrated on lexical differences between different localities, and in stylistics, which was mainly restricted to the study of style differences in literary texts. It is the merit of William Labov (1966) that he took linguistic variation as such as the central object of research, and demonstrated that phonetic, morphological and syntactic features of speech may show regular patterns of variation across a matrix defined by social class differences on the one hand, and a spectrum of formal–informal style on the other. It was further demonstrated that such linguistic differences may either receive prestige value or be negatively evaluated according to the social class parameter, and that such evaluations may trigger shifts towards the prestigious norm, which will leave their traces on the historical development of the language. Labov further demonstrated that variations across age groups may betray ongoing changes in the language, and that such changes are gradual and approximative in character rather than characterized by sudden 'jumps'.

In all this work, it was stressed that linguistic usage should be studied under natural conditions of spontaneous speech, since only in those conditions does one find out 'how people really talk', rather than how they say or pretend that they talk. Thus, the study of linguistic variation brought with it a new methodology of careful registration of natural talk in realistic circumstances, and of data analysis in terms of sophisticated statistical techniques. These various contributions have produced the powerful new research tradition of sociolinguistics.

DIMENSION V: language, languages, and Language

Position (V:1). Each language should be described in terms of its own structure.

This slogan of American descriptive linguistics arose in natural protest to the application of the categories derived from the study of the Indo-European languages to non-Indo-European, in particular American Indian languages, at the beginning of this century. The study of these languages seemed to require completely new conceptions of how a language can be organized. The fact that current linguistic methods were unable to handle them was generalized into the view that no set of pre-determined categories would be adequate for capturing the essentials of even the next language to which they were applied. In theory, position (V:1) is of course incompatible with the notion of a general linguistic theory. In practice, even those who adhered to this position contributed greatly to our understanding of the general principles underlying Language.

Position (V:2). The final aim of linguistic theory is to develop a theory of Universal Grammar sufficiently powerful to explain the properties of individual languages; much about such a universal theory can be developed through the in-depth study of even a single language.

Paradoxical though this Chomskyan position may seem, there is no doubt that the theory of transformational grammar, which was developed to a large extent through a detailed analysis of principles of English syntax, has greatly contributed to our general understanding of how any natural language can be organized. An advantage of the mono-lingual approach to linguistic theory is that one has an enormous wealth of relevant data in one's own native competence of the language in question. And clearly, if some principle of grammar is to have universal relevance, it must also belong to the central core of any one particular language.

A danger of position (V:2) is that it may lead to an ethno- or glotto-centric view of the full gamut of natural languages: properties which may seem absolutely indispensable to English grammar may turn out to be absent from other languages; properties which no one would think of when studying English may be central organizing principles of other languages.

Position (V:3). The final aim of linguistic theory is to develop a general linguistic theory sufficiently powerful to explain the properties of individual languages; such a general theory can only be developed through the typological analysis of representative samples from the full set of natural languages.

This typological position, which in fact defines a research tradition of its own, is continuously represented in linguistics from the nineteenth century, through the work of the Prague School and in particular of Edward Sapir, up to the present day. It was strongly invigorated through the 1961 conference on universals of language, in which Joseph H. Greenberg played a prominent role. Greenberg's contribution to the study of language typology was both methodological and theoretical. Methodologically, he demonstrated that significant and unexpected generalizations about language may be derived from studying a carefully chosen sample of representative languages. Theoretically, he demonstrated the productivity of the notion 'implicational universal'. These are universal statements of the general form: 'For all languages *L*, if *L* has the property *P*, then *L* has the property *Q*.' Note that such a statement says nothing about the universality of the properties *P* and *Q* as such. In fact, the universal statement defines a threefold typology of 'possible' languages: with *P* and *Q*, without *P* and *Q*, and without *P* but with *Q*. It is only languages with *P* but without *Q* that are excluded by the universal statement. Implicational universals of this type have proved a powerful instrument for describing those general properties of natural languages which determine the systematic typological variation between them. This was even strengthened when, in later work, it turned out that such implicational universals can often be connected in series of the form: 'If *P*, then *Q*; if *Q*, then *R*; if *R*, then *S* . . ., etc.' Such hierarchies have been demonstrated to be in force in almost any domain of linguistic organization. They obviously allow us to predict a great deal about the properties of a language in which we have observed the property *P*. They thus amount to bits and pieces of a general theory with predictive potential with respect to the notion 'possible natural language'.

DIMENSION VI: aims and goals

Position (VI:1). The highest aim of linguistics is to describe the languages of the world.

There was a time when position (VI:1) was quite generally held to embody the final aim of linguistics. No doubt, describing languages is a very important goal in itself. In the first place, language description often takes the form of a rescue operation: lots of languages are dying out, either through extinction of their speakers, or because they are superseded by other, mostly European languages. In the second place, descriptive data on a wide range of languages are essential to the development of insight into the general properties of natural language, at least for those who take the typological position embodied in (V:3). On the other hand, there is of course the general insight that, in order to be able to describe a phenomenon, we need some sort of general theory about that phenomenon. Description in a theoretical vacuum is simply impossible.

> Position (VI:2). The highest aim of linguistics is to arrive at an explanatory theory of the structure of natural languages.

This is the Chomskyan ideal of arriving at explanatory adequacy for the theory of grammar: '. . . my own interest has always been almost solely in the possibility of developing an explanatory theory' (Chomsky, 1982). This ideal is pursued by attempts to show that the intricate patterns found in natural languages can be derived from the interaction of a few, conceptually simple, 'deep' underlying principles. This Chomskyan ideal has certainly contributed greatly to the theoretical sophistication of linguistics. Unfortunately, in the hands of some linguists, this ideal slips into an anti-descriptive bias, according to which descriptive data are only of interest to the extent that they directly contribute to the further elaboration of this or that theoretical principle.

Perhaps it is useful, then, to recall Saussure's formulation of the aims of linguistics (1916), which clearly provides the synthesis between (VI:1) and (VI:2):

> Position (VI:3). The task of linguistics is (a) to describe all languages that can be attained, both synchronically and diachronically; (b) to discover the forces which permanently and universally determine both the synchronic functioning and the diachronic development of languages; (c) to delimit and define itself.

DIMENSION VII: language and individual psychology

Throughout its development, linguistics has been in a kind of love–hate relationship with psychology. On the one hand, it cannot be denied that language is an important component of individual psychology. On the other, it cannot be just that: no one could understand the social functioning of language if it were not, in some way, also a function of the social group. Furthermore, to the extent that language is interpreted as a psychological phenomenon, the question arises whether psychology provides the means and methods for explaining this phenomenon. A common attitude towards nineteenth-century psychological accounts of linguistic phenomena was that they explained these phenomena in terms of mental processes, the only evidence for which was provided by the linguistic phenomena themselves (cf. Bloomfield's comment on Hermann Paul, 1933).

Nevertheless, we can distinguish some influential positions with respect to this complex problem.

> Position (VII:1). Language is a psychological phenomenon, integrated into each individual mind; it is also a social phenomenon, in that the individual cannot withdraw from the linguistic conventions existing in his native society. Linguistics is part of semiotics which, in turn, is part of social psychology.

This is, very briefly, the position of Saussure with respect to this dimension. Saussure (1916) added, however, that since language is the most important semiotic system available, it is rather linguistics which will contribute to the development of semiotics than the other way round. In actual practice, Saussure makes no essential use of notions from social psychology in explaining linguistic phenomena.

> Position (VII:2) (the mechanist position). No factor may be accepted as explaining linguistic phenomena if that factor cannot be reduced to observable properties of human behaviour.

Probably through the popularity of positivistic thinking in its time, this Bloomfieldian position exerted strong influence on American linguistics of the 1940s and 1950s. There is a certain paradox in Bloomfield's work in that, on the one hand, he claimed that linguistics had nothing to do with psychology, while, on the other hand, taking such extremist views with respect to mental phenomena. The mechanist position had important consequences in that it entailed the idea that meaning must be 'somewhere out there in reality', which in turn led to the behaviourist position on form and meaning discussed above (II:2).

> Position (VII:3). Language develops as a mental organ on the basis of an innately defined cognitive programme. Linguistics is thus part of cognitive psychology.

This Chomskyan position did much to push the mechanist position off the scene. Some points worth noting about this position are the following:

(a) Chomsky's position is a strongly individualistic one: no importance whatsoever is accorded to the social side of language.
(b) It is a strongly rationalist position: the essential factors determining the acquisition of language are thought to reside in the acquiring system itself; it is almost in spite of, rather than thanks to, environmental triggers that children acquire their native language.
(c) Chomsky's conception of the cognitive status of language is not a procedural

one: the cognitive principles of grammar are thought to relate only very indirectly to what people do in actual performance.

(d) (d) In actual practice, no independent use is made of psychological results in developing linguistic theory.

We seem to see a pattern emerge with respect to this dimension: linguists either believe psychology to be immaterial to their concerns; or, if they believe that linguistic phenomena are essentially psychological in nature, this belief does not appear to influence their work as linguists in essential ways. This, however, creates an unproductive vacuum between linguistics and psychology, and it leaves psycholinguistics to its own resources for developing linguistic models with explanatory potential with respect to actual human linguistic behaviour. The psycholinguistic position is as follows:

> Position (VII:4). Psycholinguistics wishes to understand how people actually produce, comprehend, store and retrieve linguistic information; and how they acquire or lose the ability to do so. This requires linguistic principles of a naturalistic procedural nature, in terms of which actual linguistic behaviour can be explained.

Since linguistic theory, in the more limited sense of the term, does not provide such principles, there has been a growing tendency for psycholinguists to develop their own means and methods for doing the job. This is also true of artificial intelligence which, with its interest in developing realistic models of linguistic communication, appeals in vain to linguistic theory for the tools and materials needed for building such models.

DIMENSION VIII: language and social group.

We saw in position (VII:1) that Saussure conceived language as a social phenomenon. The complete linguistic system is a function of the linguistic community, which imposes itself inescapably on the individual language learner. However, Saussurean structuralism did not really lead to a sociological conception of language, because of his taking position (I:1), abstracting the system from its conditions of use.

Chomsky's position (VII:3) does not leave room for social factors either. In his case, this is further reinforced by his abstraction from any form of linguistic variation (position (IV:1)). Nor does Chomsky attach particular importance to the fact that language is a means of communication, i.e. of social interaction, witness his statement (1976) that 'communication is only one function of language, and by no means an essential one'. This effectively boils down to the following position:

Position (VIII:1). Linguistic theory need not take into account any social variable.

Several different trends, however, have contributed to positions in which more than lip-service is paid to the thesis that language is a social phenomenon. Such contributions pertain both to the micro-level of social interaction and to the macro-level of linguistic varieties correlating with social stratification.

Position (VIII:2). A language is an instrument for social interaction. It must thus be studied within the social context of inter-personal communication.

This position was taken by Dell Hymes when, in the anthropological spirit of Sapir, he suggested that the notion of grammatical competence be transformed into 'communicative competence', the ability of speakers and hearers by virtue of which they can arrive at socially meaningful forms of interaction. The position was also defended by William Labov, both in his general plea for the study of language in its social context and in his actual demonstrations of how such micro-analysis can be carried out.

The development of speech act theory and conversation analysis, as discussed under Dimension I, further contributed to the study of language as a means of communicative interaction.

Position (VIII:3). Language conventions differ according to social variables, and both determine and are determined by differences in social position and social evaluation.

This position is fundamental to sociolinguistics in the macro-sense, as discussed under position (IV:2) above. The enormous growth of this research tradition in the last decade testifies to the fruitfulness of this position.

6 SUMMARY AND CONCLUSIONS

This paper was written in a spirit of antithetic pluralism, which was recently imputed to me by way of criticism, but which I would like to accept as a title of honour. It is the spirit which, in Laudan's (1977) conception of scientific progress, is the productive force in any scientific development.

Scientific progress is characterized by the coexistence of competing and complementing research traditions. I have tried to demonstrate that this is certainly true of linguistics, and that these research traditions, in turn, can be broken down into component positions, which in different combinations define the various approaches to the common task of describing and explaining the organization of natural languages in all their manifold manifestations.

In view of the rather detailed character of this analysis, my conclusions will be brief: the coexistence of different research traditions should be evaluated as a sign of vitality rather than of immaturity. Discussions across paradigm boundaries should be stimulated rather than shunted. In teaching linguistics, care should be taken that students get acquainted with the full intellectual wealth of the field. In research, possibilities for integrating the different disciplinary and interdisciplinary contributions should be actively pursued. And in research policy, one should avoid mistaking the ship for the fleet, even if one sail may temporarily eclipse most of the others.

Comment

W. U. DRESSLER

Bei Erweiterung des Wissens macht sich von Zeit zu Zeit eine Umordnung nötig, sie geschieht meistens nach neueren Maximen, bleibt aber immer provisorisch.

J. W. von Goethe

No categorization of scientific disciplines can assign linguistics to a simple category: linguistics is a human science, but parts of it have close relations to biology (e.g. neurolinguistics) and physics (e.g. phonetics); it is a social science, but at least partially also a formal science. Due to the variety of its subject-matter and of the methodologies used, linguistics has many interdisciplinary connections (see, for example, Braga *et al.*, 1980). The most significant basis of its interdisciplinary importance lies in the essential role that language plays in all sciences of man (cf. Wittgenstein, 1958). Notice also the triumphant expansion of structuralism from linguistics to other disciplines (e.g. in France to literature, anthropology, history, sociology, philosophy and psychiatry, see Piaget, 1968) or the interest that generative grammar has found among philosophers (see Stegmüller, 1979), psychologists, anthropologists, ethnologists, etc. And from the 1950s to the 1970s linguistics has experienced a period of exponential growth.

As a consequence of these facts we find a great variety of linguistic approaches, a telling sign of the vitality of linguistics. Simon Dik's paper has amply displayed this variety of contemporary linguistic trends. I basically agree with Dik's pluralistic approach (cf. Feyerabend, 1975) and with his view that Kuhn's (1962) model of paradigm shifts is incorrect (see Percival, 1976). Dik's 'componential analysis' of current linguistic approaches and his comparison of four basic parameters is essential to an understanding of contemporary linguistics and which stages of evolution its various branches have reached.

Compare Raymond Boudon's insistence on heterogeneity in sociology. In both disciplines the plurality of approaches seems to be only partially traceable to the heterogeneity of the objects of investigation. Notice that heterogeneity within sociolinguistics is smaller than in its founding disciplines of sociology and linguistics, i.e. there is no 'compounding' of sociological and linguistic heterogeneity in sociolinguistics. This is in agreement with my contention that heterogeneity in linguistics is, at least to a significant extent, based on the variety of theories of science used (explicitly or implicitly) in linguistics.

In this contribution I want to concentrate on the issue of how to evaluate progress in linguistics. Professor Dik has laid the foundation for such an endeavour, but he has avoided investigating directly the question of 'progress' at length. In section 3 he stresses an obstacle to assessing progress, i.e. the subjectiveness of evaluating endeavours in linguistics. This is a very honest and welcome change from the all too common practice of covering only one's own school and considering its merits as self-evident.

Still, there is a possibility of balancing the advantages and disadvantages of different approaches, although the weighting of this balance may differ greatly according to the personal interests and prejudices of the evaluator. One important step in this direction is to apply a grid of criteria that has been found useful in evaluating progress in some other science. The grid I am going to apply to linguistics is the matrix used by the physicist Michael Moravcsik (1977) for assessing progress in specific branches of both theoretical and applied physics and for contributing to policy decisions in foreign aid programmes in physics.

Moravcsik's (1977) first criterion is as follows (cf. Raymond Boudon):

> a healthy sign from the point of view of progress is a close interaction between speculation and explanation on the one hand, and observation and measurement on the other.

Such a balance is sought by many linguists nowadays; earlier at times both structuralists and generativists have indulged in formalistic theorizing, whereas whole schools shied away from any attempts at explanation in favour of mere description (cf. Dik, section 5, position (V:1)). But there is a great dispute about what counts as explanation.

Moravcsik's second criterion is the cumulative nature of research in phases of great progress, in contrast to (with my emphasis):

> a field of science . . . characterized by many different approaches, speculations, conjectures, models, and experimental directions, *unrelated to each other*.

Thus, so long as different approaches profit from each other's result, progress is more probable than in phases where different schools coexist in splendid isolation. According to this criterion 'hyphen disciplines' such as sociolinguistics, psycholinguistics, etc., show healthier signs of progress than many brands of theoretical linguistics.

Moravcsik's third criterion stresses 'new conceptual idea[s]' and 'novel ways of looking at a problem'. Among the large linguistic schools generative grammar excels in this respect; however, valid innovations can be consistently distinguished from trendy fads only if Moravcsik's first criterion is applied simultaneously.

However, innovations cannot result in lasting progress if they are achieved at the expense of valuable traditional standards. One case in point is the genesis of neurolinguistics: in order to study the dependence of language on brain, be it in performance of speech, in language acquisition or in language disturbances, neurolinguists have adopted neurological and psychological research paradigms with great enthusiasm, but they have neglected time-proven philological and linguistic standards of data collection, of data analysis, and of data display in publications (Dressler, 1983). This impedes lasting progress.

Moravcsik's fourth criterion is the degree of mathematization. This criterion may be very important for physics; in linguistics, quantification has brought about progress only in a few areas, such as linguistic typology and computer linguistics, and by the use of statistics in certain areas of applied linguistics, phonetics, psycho- and socio-linguistics. Elsewhere, veneration of mathematics has led linguists into many blind alleys, because linguistics is not a natural science, but a conceptual field (cf. René Thom).

Moravcsik's fifth criterion connects progress with 'an increasingly broader unification of our view of nature'. Here research in universals (see Dik, section 5 positions (V:2) and (V:3)), the quest for valid generalizations in generative grammar immediately come to mind, and the enormous extension of the scope of linguistics in many approaches.

The sixth criterion is 'further achievement of economy and simplicity of explanations'. Economy and simplicity are quite controversial issues among schools (see Grassl, 1979, and Bailey, 1982).

Moravcsik's 'seventh methodological criterion for scientific progress is applicability'. In this respect various branches of linguistics have made enormous progress: for example, linguistics has been able to help in making laws comprehensible; in detecting misunderstandings and dysfunctional communication between medical doctors and their patients; in helping to diagnose pathological language and in aiding speech-pathologists for devising therapies; in helping police to catch criminals by analysing telephone calls of criminals; in evaluating terminologies and in proposing improvements in normative terminology. Typically such achievements have been based on interdisciplinary methodologies including sociolinguistic, psycholinguistic and textlinguistic research paradigms.

Last but not least, let me touch upon the thorny problem of how to decide which linguistic projects should be financially supported:

(a) In most respects linguistics is a cheap science.

(b) Projects in purely theoretical linguistics rarely need much money, and current evaluation procedures seem to work no worse than in other disciplines.

(c) Projects in applied linguistics may cost more money than is usually awarded to linguistic projects. Here I would like to make the following tentative recommendations:

  (i) All those projects which demand considerable extra funding should be of an interdisciplinary nature, such that linguistics and the other disciplines needed are firmly integrated.

  (ii) For a trial run, only relatively cheap pilot projects should be financed; only afterwards should more expensive long-term projects be supported.

  (iii) There are many areas of relevance to society at large where linguistics could make a significant contribution, and where it should be included in an interdisciplinary effort.

How is scientific progress assessed in art history?

G. KAUFFMANN

1 PRELIMINARY REMARKS

AN approach to the problem of assessment of scientific advancement in art history is only possible within the wider framework of a general discussion of progress in modern science. As background literature I cite the following publication, *Die Philosophie und die Frage nach dem Fortschritt*, edited by Helmut Kuhn and Franz Wiedmann (1964).

2 PROGRESS IN ART HISTORY

2.1 Progress as movement

The idea of movement is at the basis of the concept of progress. Since movement can only be observed by comparison with something stationary, it is necessary in order to define movement to identify the constants associated with it. In history of art there are two fundamental constants: first, the *status quo* of scientific knowledge, the state of research, and secondly, the object of scientific investigation, the work of art.

(A) *Progress in relation to the status quo, or to the state of research.* To speak of progress, it is necessary to have a clear idea of the *status quo* from which it will evolve. This *status quo* is the state of research at a given time. The appearance of each new stock of 'monuments' may be taken as progress, in comparison with the present state of knowledge. For example, the work of Wilhelm Lübke on the art of the Middle Ages in Westphalia (1853) considered for the first time the great monuments of Westphalian architecture of the eleventh to thirteenth centuries, not only in individual monographs, but also, more comprehensively, as representative of a whole group, thus enriching and refining what was then known about German Romanesque art. Some 100 years later, Kurt Wilhelm-Kästner published *Der Raum Westfalen* (1955). At the very beginning, the author referred to the literature which had appeared since Lübke; thus, presenting a profile of the state of research, he could perceive his own work as progress, as a step beyond what had been accomplished before. Progress in this instance is characterized as 'breaking away from . . . ', 'going beyond . . .', 'rising above . . . ' (specifically, above the *status quo*).

This departure in one's own work from established knowledge confirms that science is discussion, dialogue. Through this explicit or implicit dialogue, science provides an insight into the mechanism of the drafting phase, in itself centred on progress, through which the exposure of this reflection on the state of research has also a didactic intention which serves the better dissemination of science.

(B) *Progress in relation to the object, i.e. the work of art.* We note that today fewer authors take the trouble to refer to the state of research in order to define the basis from which they start; science thus becomes a monologue, occasionally a positive profession of faith linked with another type of progress. Today science does not seek to rise above the present state of research, as it is attracted by the object, so that we can consider progress as 'getting closer to . . .', 'leading towards . . . ' the work of art. In this way, it is not so much science, but rather the existentiality of the work of art, which is the centre of interest. For instance, Oskar Bätschmann, *Dialektik der Malerie von Nicolas Poussin* (1982). In the preface, the author explains that he looks straight at works of art and does not try to build on the basis of coherent research.

(C) *Comparison.* If we compare these two types of progress considered as movement – which never occur independently – we can describe them as follows:
Type A (Progress in relation to the state of research). It rests on a positivistic as well as a philological basis. It reached its climax when emphasis was being given to increasing our knowledge of monuments. As time went on the three following risks appeared:

- With the discovery that an actual store of monuments was coming to an end and there was an absence of 'white areas' on the map, scientific work of this type could only consist in correcting erroneous opinions or elaborating earlier statements. Progress then falls into a secondary realm, as it leads only to views deduced within the framework of wider relations which have already been defined.
- Whoever tries to discover additional new territory can only find land of lesser quality. Originally, art history focussed on the 'key' works and the central questions characteristic of a period. It had developed its methodology through interactions with them and had established the leading principles to which the mass of average pieces were generally subjected. Consequently, it can be considered that investigation of average work means falling to a secondary level. These studies lack, from the start, the possibility of yielding really new results,

hence their not uncommon feature as enumerations or inventories. A recent example is Henry Russell Hitchcock's *German Renaissance Architecture* (1981).
- Art history is to be understood as the science of monuments. The monuments allowed us to name the discipline in those terms. If it makes way for the unimportant, it takes this well-established concept and applies the concept of monuments to mediocrity. But in so doing it adds mobility to the average and at the same time encourages inflationary devaluation of the concept of monuments; progress results in a levelling of the scale of values. We are facing a typical problem of our time, not to mention the practical problem of preservation *Denkmalpflege*.

To these three cases we may talk of movement, but not of real progress. In fact, we are dealing with the illusion of progress. This illusion appears as a necessary consequence of scientific activity, with a negative content founded on the logic of scientific behaviour.
Type B (Progress in relation to the work of art). It is of a philosophical nature. It is not the mechanism of scientific methodology which is decisive, but the uniqueness, at the time, of the work of art. What is at stake is not the discovery of the material but its intellectual penetration. The researcher can develop his activity with more freedom and can inject more of his inventive faculty into it. Only then he becomes an 'interpreter'. Since Sedlmayr differentiated between 'primary' art history – by which he meant the discovery of the material – and what he called 'secondary' art history ('Zu einer strengen Kunstwissenschaft', in *Kunstwissenschaftliche Forschungen*, 1931), the latter has been reserved for those who believe themselves to be the better art historians. It is quite frequent for outsiders to practise it as well. In this case progress must be assessed differently. There is no obvious platform for a state of research.

Since the work of art as the central point of reference can be the object of different individual disciplines, the possibility of an interdisciplinary approach arises. If that approach cannot offer something satisfying, this is due to the illusion that a common object of study is sufficient for a common activity. This common object looks different to every discipline and with each special field using its own methodological approach, the results can be difficult (or even impossible) to co-ordinate. It is therefore not primarily the fact of co-operation between disciplines which provides some hope of progress, but rather the combination of disciplinary methods (which might have to rely on preparatory steps).

Here the humanities come closer to the natural sciences, which recognized earlier that each methodological instrumentation contains its own potential for discovery. With an interdisciplinary approach, progress in knowledge can be achieved only when the methodological instruments of one discipline can be transferred to another. Knowledge can be defined as connected with instrumentation and possible to achieve only within specific methodological operations. Thus, it is the phenomenon of transferability of methods which guarantees the success of interdisciplinary activity (Kah Kyung Cho, 'Naturbild und Lebenswelt', in *Denken und Undenken – Zu Werk und Wirkung von Werner Heisenberg*, 1977).

But dangers lurk here too. An exclusive focus on the object can result in losing sight of the context in the process of work carried out by individual researchers, because their results cannot be projected onto the common base of the state of research. Our expert congresses today look more like a series of isolated contributions, the connection of which is sought through the formulation of a common topic but is rarely achieved. What we lack is the integration and co-ordination of knowledge. As a consequence, the transparence is blurred. It becomes difficult to identify the essential amongst so abundant a production. One no longer sees what is really new. It becomes very difficult to distinguish between real advancement and the mere semblance of progress resulting from innovation. Innovation claims to be considered as an absolute value. Real progress, however, is relative, in the present case relative to its supposedly measurable advance towards matter itself.

2.2 Progress as a process

It is possible to substitute the concept of transformation to the concept of movement. Progress is then experienced as a process. In this context, it is irrelevant to decide whether we mean that the discovery of new facts, which in the end can no longer be explained by an existing paradigmatic scheme, leads to a crisis which can be resolved only by the revolutionary introduction of a new paradigmatic scheme; or whether we tend to adopt the opposite view that it is not new facts which bring progress but a changing point of view towards existing knowledge (cf. the reports of the Meeting of the Society for Science History, Münster 1965).

(A) *Logic of the construction of genres and classes.* While the more we think we know a field, the more we can subdivide it, scientific progress appears in general to be something which can classify, specify and create genres. Nowadays, the role of the historian seems more than ever to be considered as that of condensing, with the help of various systems of classification, the enormous number of 'signals' which reach us from the depths of history. The systems result only partially from the culturally determined classification of the objects under investigation; they are to a high degree the product of scientific logic. Among the classes which were really created by art history, one can mention Cistercian architecture or devotional pictures. There have been quite refined subdivisions in the Dutch painting of the seventeenth century, with a distinction between still lifes, portraits, flowers, historical paintings and landscapes; this is also true of graphism where, for instance, very subtle differences have been introduced (cf. B. Degenhart, 'Autonome Zeichnungen bei mittelalterlichen Künstlern', in *Münchner Jahrbuch der Bildenden Kunst*, 1950). Here we run the risk that the logic of the scientific construction of classes would not do justice to reality. The *problématique* of classification arises from the simple fact that new things appear regularly which, at first glance resist the specification of genres. This has recently been applied by Sandra Hindman to the history of books (Sandra Hindman and James Douglas Farquhar, *Pen to Press, Illustrated Manuscripts and Printed Books in the First Century of Printing*, 1977). It is obvious that there are major differences between the old handwritten manuscripts and the printed book of modern times, all the way to the organization of the text itself. But science has underlined these differences too strongly and given them, through the work of individual disciplines, a fundamentally disrupting, absolutely distinctive character. Each specialized field stakes its own claim, like gold-diggers, and consequently historians reserved the printed texts, art historians the illumination of manuscripts whose painters, styles and iconographic sources became the exclusive object of their investigations. The activities of these two research trends reinforced the feeling that there were two totally different territories and the prospect of a common synopsis and overview became all the more bleak as each one stuck more closely to the methodology of its own discipline. But it is only when both manuscripts and printed material are considered from the same perspective that one becomes aware of the continuous transition. When Hindman looks at these mixed products (so-called pseudo-manuscripts or hybrid-prints), she is aware of the fluctuations between manuscripts and printed material. Looking at these pieces, the power of classification to define is diminished.

In the work of Hindman and Farquhar, progress is achieved against the tendency inherent in science. The weakness of classification is made obvious in this research achievement. It is not rare to see new creative forces break through precisely in transition phases where classification in genres becomes difficult: does this mean we should refrain from relying on genres precisely at very creative moments in history? Moreover, the question of genres includes that of the actual meaning of the object, since genres occupy a stronghold in the cultural structure. Too bad for pieces which hesitate between genres. They cannot find a place in the scale of values, and therefore the product of transition is devalued as indicated by the use of devaluating concepts such as pseudo-manuscripts or hybrid-prints. To think in term of genres turns the unconventional into the abnormal.

(B) *The sophistication of the paradigmatic framework.* We are now at a point in the history of the humanities when well-established basic principles are being questioned. This can be the result of pure complication. One of the basic principles of art history (referred to here as 'the paradigmatic framework') is description. One can say that, in the course of scientific progress, describing has become more and more difficult.

Description paves the way which leads us closer to the work of art. Describing develops from observing. But observing – as we know now – is far from establishing objective hard facts, since the manner of observing will determine which features of the work of art are enhanced and which left aside or eliminated. Observation thus works in two ways: on the one hand, it establishes absolute, active interaction between an object and an observation system based on scientific instrumentation, an interaction independent of the subjectivity of the observer; on the other hand, observation as an act of perception is a recording of results which draw on this observation system. This double function is the decisive cause of contradictory interpretations. We consider now (as distinct from what we thought earlier) that it is impossible to reach *the* interpretation which would be authoritative. Such an interpretation can in any case not hold good, for the simple reason that everything we establish is in the realm of possibility only. One can only interpret a work of art by describing the possibilities – in the sense of tendencies – which apply objectively to it under specific macro-conditions. If one grasps the actual, recognized condition of a work of art as a simple possibility, then the concept of 'coexisting' states becomes plausible, as one possibility does not necessarily exclude another.

Curators and restorers nowadays do not shy from making such coexisting states visible. The new industrial archaeology turns technical buildings of the previous century into cultural monuments, because coexisting with our world today, they provide documentation on an older social order. We can see from the Gedächtniskirche in Berlin, the Notke Triumphkreuz in Lübeck, the Klarenaltar in the Cologne Cathedral (to mention only a few) how different eras can exist beside each other. For the naive observer, the work of art therefore loses its uniqueness (or identity) which can provide him with access through experience.

Obviously, we can draw a conclusion here: the more complicated the paradigmatic scheme, the more difficult the access through experience.

2.3 Progress as a problem

(A) *Lack of visualization.* One of the results of progress in modern science is the growing lack of visualization. In art history as well each instrumental logification leads to abstractions. Abstractions may be defined as a removal from historical reality. They alter this historical reality because they contribute to the elimination of entities or, to be more precise, to the elimination of problems by reducing some entities to others, as is also the case in the natural sciences (for example, reducing 'sickness' to chemistry, chemistry to physics, physics to mathematics, mathematics to logic). Such reductions can be considered as progress insofar as they produce interesting results, but at the same time they bring a loss of reality. Such procedures are only made possible through the fact that the expression of scientific theories, as a matter of principle, goes far beyond the data resulting from observation.

Subordinated processes accompany each interpretation. The error therefore lies in the false as well as the naive belief that there are direct correlations between ontological and empirical facts. Why otherwise would the sciences have developed their own systems of signs? A chemical formula contains nothing of what we learn from the sensuous observation of the material in question. Instead, the formula places the material in a well-articulated system of relations, of which the observer is not aware. This is also true of art history which does not grasp its objects according to what empirically 'is', but as the embodiment of possible relations, reactions and correlations to be explained by causality. Since science and visual perception began to part in the eighteenth century – this has been shown, for instance, by E. Panofsky and J. Ackerman – the problem arises, specific to science, of how to go back to intuition. We cannot recognize as progress what does not first lead to abstractions, but then leads back, with the same assurance, to the object as a whole and to the recording of real facts.

(B) *The necessity of intuitions.* Consequently, there is a phenomenon which is complementary to the sophisticated abstraction processes, namely the appearance in art history today of a tendency to reintroduce pre-scientific, even naive, perception of art. As art and science have never been in total harmony, without friction, art history has always allowed space for the assistance of intuition. At present it does so with special emphasis (cf. the title of the commemoration volume for Hanns Swarzenski, *Intuition und Kunstwissenschaft*, 1973, as well as H. Lützeler, *Kunsterfahrung und Kunstwissenschaft*, 1975). Our research again stresses non-scientific attitudes towards art. Besides the rational, we must have the irrational; besides the history-conscious we must have the a-historical, so that progress can also consist of removing the unfathomable from a spontaneous experience, while respecting it and keeping it intact.

Again, it is a matter of 'abstractions and empathy' (Worringer). Both should complement each other in the nascent vocational definition of museum pedagogues. Everyone can have access to art culture. Therefore science should not be considered as having secondary importance. To say that science bars artistic experience is a prejudice, which obviously hides the ineradicable need 'to save the immediate truth of the sense-impression against the offensive of science' (Helmholtz when characterizing Goethe's theory of colours).

3 EXTERNAL AND INTERNAL FACTORS OF PROGRESS

Progress effects changes. It changes the idea of a discipline, it changes the discipline itself. Both external and internal factors play a role in the process. External stimulations gain significance to the same measure as internal strength gives way. They hide the danger of blurring disciplinary boundaries. They also mislead the scholar with the temptation of loosening methodological stringency.

3.1 External factors

(A) *Money*. Well-founded research in art history necessitates only a little financing. If the offer of funding from outside increases – which is not so rare – one is pushed into devising projects which would otherwise (and preferably) not have been born.

(B) *Politics*. Research impulse can be generated by the expectations of politicians, without being really followed up. Only those questions which arise from science itself prove to be lively. For instance, the concept of 'cultural monuments', which created great hope in certain parliaments, has not yet been properly defined, although there have been promising attempts.

(C) *Marketing*. The market demands reproduction. In this respect, the situation of art historians can be compared to that of practising musicians (cf. European colloquium in Strasbourg, 25–26 November 1982 on 'Music and the industrialization of culture in Europe'). There is an extensive production of art books with colour plates of the highest quality. There exists a general commercial tendency to produce bestsellers, often at the cost of a higher scientific *niveau*. It is true that this sector of the culture industry does not contribute to the promotion of scientific creativity; it does, however, have a significant effect on the state of general culture. Rigorous scientific endeavour should not ignore this phenomenon. Here science is the object of processes of differentiation by which the ability of art history to 'go public' increases. The following from H. Lübbe's

'Kunstwissenschaft und Kunstinteresse – über kulturpolitische Folgen des sozialen Wandels' (1983):

> Culture-political complaints about the occasional failure of science to reach the public do not stem, therefore, as far as cultural history is concerned, from the idea that these sciences cultivate an unnecessary esoterism. The problem lies more in the fact that it is today more difficult than ever before to transpose the necessary esoterism into exoteric information, while at the same time the call for information of this kind in the public is stronger than ever.

The external factors which have an impact on art history are countless. The totality of the complementary interrelations between developmental dynamics and historification in the cultural environment significantly affects the present situation of our science and in this way the interrelations characterize progress.

3.2 Internal factors

The number of internal stimulations is also of course legion. I will mention just two examples.

(A) *Teaching.* This is today, in art history, the most important guardian of the transmission of our scientific tradition. It does not make sense to want to provide the new generation with only the present state of research, since they would not understand it without having a sound knowledge of its foundations. To talk about Gothic style, mentioning only Branner without having explained first about Vöpe and Jantzen is just as fruitless as not mentioning problems of epoch and style just because they have been studied mainly by the previous generation.

(B) *Team-work.* There is sufficient evidence that collaborative research produces good progress. However, the driving force of scientific progress in art history remains the individual working on his own. The employment of collaborators remains problematic (see Niklas Luhman, in *Forschung in der Bundesrepublik Deutschland*, 1983).

4 TWO NEW ELEMENTS

Scientific progress in art history seems to be characterized today by two new features, the tendency to historical distance and the idea of progress as change.

4.1 The tendency to historical distance

While in earlier times an effort was made to 'understand' the cultural past through sympathizing (through identification with it), today – following the tendency in the study of literature, and with the explicit support of some representatives of classical archaeology – there is a demand for a certain distance from historical monuments. A. H. Barbein made the following comment in his paper 'Archäologie und historisches Bewußtsein' in 1981:

> Historical awareness is to be achieved anew through a constant process and has a much wider-reaching meaning than mere historical interest. The experience of distance is one of the most important prerequisites: it is only when the different, even strange character of the past is grasped and accepted that concern and participation play their real role and that communication can take place, overcoming the distance.

That also the museum has the task of promoting not identification with the works of art exhibited, but on the contrary creating distance, is maintained by Norbert Kunish in Jahrbuch 1982 of the Ruhr University in Bochum.

To be precise, this tendency to keep one's distance is an antihumanistic tendency. Here progress is a loss. From other points of view, too, art history does not always appear to be a humanistic discipline. It is part of this assertion that, in recent times, one feels the need to replace the concept of humanities by the more neutral one of 'historical culture studies'.

4.2 Progress as change

The idea of progress is part of any modern science. In fact, it has the head of Janus. Seen next to progress as a free creation of the spirit, progress also appears as a forced development. That does not always lead to evolution. It can be a matter of pure joy in change; the best examples are still those which bring to the eyes (and beyond the stifling scientific ethic, with its burden of problems) something of the happy irresponsibility of the *aperçu*, 'the scant beauty of which goes just under the skin but not more deeply' (A. Gehlen). This kind of scientific movement has nothing to do with progress. It is having a good time within a stationary general situation.

Comment

I. JONSSON

My reading of Professor Kauffmann's interesting paper has raised a number of questions which are provocative in more than one respect. Since I am no art historian myself, you must let me take my examples from the history of literary scholarship, and I think you will find that most of what is said in the paper seems applicable also to my discipline, comparative literature.

For some time during the reading I have been in doubt whether the idea of progress may be relevant or not for the humanities. An attempt to fight against such doubts could perhaps take the form of a high-speed repetition of the history of humanistic learning. As humanists we represent disciplines as old as western civilization itself – which of course is true of many scientists too – and let me start with a famous quotation from Whitehead: 'The safest general characterization of the history of Western thought is that it consists in a series of footnotes to Plato.'

As a starting-point for a discussion of the idea of progress such a statement is somewhat depressing, but nevertheless learned men, and as time goes by women too, within the humanities have seldom displayed any real doubts about the progress of learning, at least their own. When the *studia humanitatis* were established in the Renaissance, rhetoric and philology were their main assets, and manuscript studies and editorial work have always presented many opportunities to vehement controversies as well as sturdy criteria of progress. In this interdisciplinary context I may be allowed to remind you that most branches of science have a common origin in ancient texts: even Copernicus was the editor of a Greek manuscript.

But the unity of learning dissolved successively, alas, from the seventeenth century onwards, and 'the dissociation of sensibility', to quote T. S. Eliot, which started then, resulted in a chasm not only between art and science but also between the humanities and natural science. In the romantic era starting in the last decades of the eighteenth century humanistic scholarship emerged in its modern shape. History became the fundament of all studies of human culture and developed a set of critical methods aiming at establishing 'wie es eigentlich gewesen', the study of literature and the other arts withdrew from its classicist basis and found a new fundament in the idealistic aesthetics of German transcendentalism, the comparative study of languages grew in the wake of what the French call *la renaissance orientale* and developed into one of the greatest achievements of nineteenth-century learning. At the same time the foundation of our technical civilization was laid by a number of important discoveries by Ørsted, Faraday and others. By and by Romanticism was driven from the literary throne, and in the second half of the nineteenth century conquering science invaded both literature and literary scholarship as a paradigm. When Emile Zola wrote a little book about the experimental novel in the middle of the 1880s, he referred to a book by Claude Bernard, the great physiologist, as his model, *Introduction à la médecine expérimentale*. The prominent theorist of naturalism, Taine, had then already drawn up a programme for the history of literature, which implied three main aspects of the investigations, 'race, milieu et moment', i.e. elements within reach for methods related to or transferred from science.

Let me quote a statement made at the same time by our great Swedish literature historian Henrik Schück, which contains an attitude representative of positivist thinking in the humanities: 'History of literature has no right to act as judge. Its only duty is to establish facts.' There is no doubt about it that Schück here looked upon himself as a spokesman of progress, and in the actual historical situation I think it was a legitimate pride. His programme was directed against the remains of Romantic aesthetic speculation, and it implied a sound respect for incontrovertible facts, which made it possible to separate what is objectively ascertainable from mere subjective interpretations. I sometimes feel envious of the assured self-consciousness of these positivist scholars, and particularly of the almost unlimited access to virgin soil which was theirs. They could also be certain of a widespread interest in their reports from the part of the middle class public, out of which they themselves emerged.

Most of that has been changed in our tragic century, and the seeds of this change were present already in the early fruits of positivism. In his essay *Die Einbildungskraft des Dichters* (1887) Wilhelm Dilthey drew a very sombre picture of contemporary aesthetic thinking. Both the formalist poetics, which had its roots in Aristotle, and its conqueror, the German idealistic aesthetics of contents, had lost their attraction and had been replaced by anarchy and misology, a profound and fierce distrust of the philosophy of art from the part of the artists and their audience: another version of the widely spread prejudice which Professor Kauffmann quotes from Helmholtz.

As things were, Dilthey could not accept the positivist attitude of the humanities either but was looking for new ways of establishing their distinctive character as well as their justification. He found a solution in the well-known dichotomy *Erklären* and *Verstehen*, explanation and understanding: science aims at explaining natural phenomena by determining their causality; the humanities, in which Dilthey included theology, law and social sciences as well as traditional *humaniora*, are trying to reach an understanding of cultural processes which are essentially unique, not repetitive.

Dilthey's pair of opposites is obviously too rigid in both directions but nevertheless it has been of great importance for the development of modern hermeneutics, which is a basis of recent debates on the character and function of the humanities. In the first place it started many attacks of the positivist paradigm. Scholars like Schück were accused of devoting their energy to anything but what should be the primary task of literary study, i.e. to examine literary texts as artistic specimens. Instead of performing this duty, the study of literature focussed on a number of secondary prerequisites of the texts, historical framework, biographical circumstances, social background, etc. Even if this kind of censure has often been unfair and exaggerated, it was certainly not unfounded.

Of course, you cannot summarize the twentieth century development in a simple formula, but it has no doubt placed the literary texts as autonomous pieces of art much nearer to the centre. This process might be described as a movement from history and philology towards philosophy and linguistics. To a great extent this is a consequence of the development of literature itself, particularly the rise and triumph of lyrical modernism. This kind of poetry is often highly esoteric, as you all know, and thus the texts offer real challenges to scholars with penetrating and sensitive but also imaginative minds. Modernism being an off-spring of nineteenth century symbolism preserves many of the characteristics of German Romanticism. It is only logical that recent reactions to new criticism in the USA and Europe, which you may have met in the shape of 'deconstructivism', are attacking these nowadays older critics as spokesmen of romantic ideas, especially their belief in the particular power of expression with which the poetic symbol has been credited since Goethe. These present attacks have reinforced current conceptions of the autonomy of poetry, so that they sometimes seem to come close to isolation: the texts are deprived of all references to a non-linguistic reality.

This means that the modern study of literature has its own share of the same complications which Professor Kauffmann reports from his field, among them the problem of defining criteria for scientific progress. Even if you may have access to the author's intentions, when he or she wrote a poem or a novel, every interpretation will be determined by the interpreter's individual experience and his or her *Erwartungshorizont* to a certain extent. In later years the resurrection of authorities like Marx, Nietzsche and above all Freud has inspired several variants of what Paul Ricoeur has called 'the hermeneutics of distrust', the import of which is that the interpreter must try to penetrate the conscious intentionality and reach the intentions of the text itself. That may sound like a paradox, but a feeling of the words taking command is not unfamiliar to anyone writing a personal letter. But of course the problem is to make a distinction between plausible readings and interpretations that are too imaginative or even private. We have to face facts and admit that the hermeneutic revival, however stimulating it may have been to humanists, has made room for uncontrollable or even crazy interpretations. A perspectivistic theory like that of Wellek and Warren can, of course, be attached to a high degree of tolerance, but to me it is an indispensable condition that every interpretation must keep its close contact with the text and always take the whole context into consideration.

Even so it will be impossible to exclude all subjective elements, when you have to make up your mind whether you have met a real talent with a creative mind or a mere echo of the latest news from Yale or Paris, i.e. discriminate real progress from one more case of *variatio delectat*. The situation is only too familiar to anyone who has served as a member of committees selecting applicants for professorships, and the thought of potential mistakes is quite painful, especially in Sweden where your written proposal is a public document available to anyone. But there is nothing else to do but to make use of your own experience, and I think that scholars should assert themselves as experts in literature too, even if anyone can read a book.

Having spoken so much about the difficulties of interpretation I do not want to give the impression that no one is interested in the history of literature any more. That is definitely not the case, and I would like to give you one example, which may also illustrate the problem of assessing progress. One of the classical problems of literary history is the so-called Homeric question. According to ancient tradition, the Iliad and the Odyssey were both written by Homer, a poet endowed with a number of more or less legendary attributes. The modern discussion started towards the end of the eighteenth century, when the famous German philologist F. A. Wolf claimed that Homer could not possibly be the creator of these enormous epics, since the art of writing had not yet been invented in his time and no one could memorize such an amount of poetry. Wolf thought that there must have existed a number of smaller epic episodes which were transferred orally and had later on been edited as two great poetic narratives.

The argument has lasted for almost 200 years now and forms one of the biggest interdisciplinary efforts in the history of humanities, to use an expression by Wolfgang Schadewaldt. Many brilliant hypotheses have been framed and abandoned *pari passu* with increasing archaeological and philological knowledge, including the discovery that the art of writing had been invented much earlier than in the eighth century B.C. The outcome of the discussion seems to be that for the time being we are back at the ancient tradition from which it started. If you expect definite answers to clearly defined questions from a scholarly effort you will find it difficult to call the Homeric argument a piece of progress. But if anyone might be prepared to look upon it as a waste of time and energy, I believe that he or she is making a mistake. For the discussion has undoubtedly given us a much greater understanding of the Homeric epics as works of art, and furthermore it has helped to keep them alive as essential parts of Western tradition. To me it is a matter of course that the scientific study of literature has one of its most important functions in serving as the literary memory of society.

Because of that it may be legitimate to evaluate scientific results in terms of progress, even if they should mainly consist of a restitution of knowledge that has fallen into oblivion. One of the most influential books in modern literary history, Ernst Robert Curtius' *Europäische Literatur und lateinisches Mittelalter* (1948), restored rhetoric to its proper place in medieval literature, and thereby it gave twentieth-century readers a new chance of approaching the texts as they were intended to be read. I think that Curtius' contribution might be an example of how valuable a perspective of *Alterität* – historical distance – can be. We have learned from him that we should never expect any realistic representations of sensual impressions from medieval writers but look for the actual application of rhetorical patterns even in contexts that seem to be describing real events. This is not an anti-humanistic attitude to me, but rather a way to a deeper understanding of a stage in the development of literary art, i.e. an important part of the humanistic heritage.

Let me add some concluding remarks elicited by Professor Kauffmann's strictly negative opinion about political research impulses. In the Swedish debate of the government's research policy the so-called social relevance of scientific projects has filled a large place. My impression is nevertheless that social aspects always come second to internal scientific estimations, as far as the humanities are concerned, at least in our national scientific council and other research funds. I may be wrong, and of course our long democratic tradition in Sweden has saved us from many painful experiences of disastrous political interference. Still, I am not particularly afraid of any political steering of research in our open society, and I am even prepared to consider some political impulses as useful. At least they may be token of an interest in what we are doing, which has come as a surprise to many of my colleagues in later years.

I will give one concluding example: at the end of the 1960s we had one of our many school reforms, and then a government committee made a proposal that the teachers of Swedish should be given some training in literature written for children and young people. At that time there had been very little research activity in that field, but after 15 years it has proved itself quite attractive and productive, and we have just got our first professorship in children's literature at my own institute in Stockholm. It goes without saying that we applied the same standards for choosing the holder of that chair as always. The primary responsibility of a scholar must be the advancement of his or her branch of science, and no one should compromise with its claims. But we are all citizens responsible for the well-being of our societies, and I find it hard to believe that the ivory tower is an adequate place for the study of *la condition humaine*.

The identification of scientific advancement: a contribution on history

H. L. WESSELING

What I have to offer is not to be compared with the learned and well-prepared papers that have been presented to this colloquium. It is just a few reflections that came to mind during the discussions, some modest remarks on a vast and ambiguous topic, the identification of scientific advancement in history. The first problem is that 'history' itself is an ambiguous word. As far as I know, history is the only science – if it is a science (I shall come to that later) – that has only one word for the subject itself and the study of it, a curious situation, so nicely illustrated in the novel *Lucky Jim* by Kingsley Amis, in which the secretary of the professor of history used to answer the telephone with the words 'History speaking'. Obviously, in the context of this colloquium, the word history refers not to the subject-matter but to the study of it. So there is no problem there. But if there is no problem in using words like 'the study of history' and 'the writing of history', this does not necessarily imply that we can also speak about the science of history. Here, two problems come to the fore: first, is there one science of history or should we rather speak about the historical sciences; secondly, is history a science at all? Let me disappoint you again by telling you that I shall skip the second, difficult question, because that is a very complicated and indeed philosophical problem which would take us far away from the subject of our colloquium. I shall limit myself to the first question which brings us to the problem of the division of labour which has been discussed in the other papers. Then, still following the lines of our discussion, I shall come to the history of the discipline, before saying a few words on the assessment of scientific advancement itself.

It is already obvious from the programme that there must be more than one science of history, because there was also a paper on art history, a field which could logically be considered a subdiscipline of history, but is obviously viewed as a discipline in its own right. One does not ask a historian to speak on art history, one asks an art historian. Now, if you were also to have papers on the history of literature, or the history of philosophy, or the history of science, you would have a similar situation. These specializations within the field of history are not taught in the departments of history. People who specialize in these fields have had a training in, say, literature or in science before becoming historians and although, logically, they all study aspects of the subject-matter of history, in fact you will find them scattered in different faculties and schools and they have a different background.

There is another division of labour, which is even more debatable and results only from reasons of practicality. Chinese history, for example, should really be considered as just another field within history. But in actual practice, a historian of China is often called a sinologist and he or she frequently works in the department of Chinese studies, not in that of history. This has nothing to do with problems of methodology, but is just a matter of practicality: Chinese is more difficult than history. Having in this way rather drastically reduced the subject-matter of history to what is considered history in the normal sense of the word – that is roughly speaking to western history – there are still more subdivisions to be made. For the problem of the division of labour it is useful to give you a very brief overview of the way specializations in history are organized.

When I was listening to Professor Ziman's key-note speech, I was not only very impressed but also very jealous of the beautiful charts and graphs he showed us to illustrate how specializations in the field of physics are organized. I was jealous because I didn't have anything as beautiful to show you, but at the same time I realized that these two-dimensional graphs would not apply to history, because specialization in history is organized at least along three axes – of time, of space and of theme or topic. So I should need a box to show you where a specialized historian would find himself within the scientific community. Let me briefly explain this.

The most traditional of divisions is into periods, which is essentially an invention of the nineteenth century. History is broken down into ancient history, medieval history, modern and contemporary history. These are all fairly arbitrary divisions. Ancient history is roughly speaking the history of the Mediterranean world until A.D. 500; medieval history consists of European history from A.D. 500 to 1500; modern and contemporary history are already more difficult concepts, because in different countries they have different meanings. In France, *histoire contemporaire* means roughly 'since the French Revolution', whereas in the Anglo-Saxon world 'contemporary history' means 'the history of the twentieth century'.

The second division of history is according to the borders of certain regions, of geographical lines. We should realize that history, as it was founded in the nineteenth century, was first and foremost the history of the fatherland. In Leiden we still have – and I think it is the only one in the world – a chair that is officially called 'the Chair of Fatherlandic History': it is not Dutch history, not national history, but fatherlandic history. In Holland, next to fatherlandic history we have general history which means the history of other countries. In the United States, specialization is strongly directed towards national histories. There you have French historians, Italian historians, German historians, and so on, and indeed it is much more likely that in the United States the historian of China will work in a history department. This is not so much the case in Europe. Still later a third division came into being, one according to subjects. This was due to the emancipation of certain aspects of history, such as economic history and, later on, social history, agrarian history and so on.

If we have three axes according to which specializations are defined, my impression is that periodization is still the most valid and the strongest one. Historians very exceptionally move from one period to another (although we should remember that Arnold Toynbee was trained as an ancient historian before becoming Director of the Institute of International Affairs and thus a historian of the most recent history one can imagine). These things happen, but very exceptionally. If and when historians move from one period to another, it is always forwards and never backwards: a medievalist may become a modernist, but you will never see the opposite happen.

Specialization according to countries is not very important in Europe. My impression is that in the United States there is a very strong division of labour along these lines, probably for career reasons. One starts one's career as a French historian and will, generally speaking, be a French historian for the rest of one's life.

The third form of specialization, according to themes, subject-matter, and certain aspects of history, such as social, economic and diplomatic history, seems to be the least important and also the one most subject to debate. In the first place, it is very hard to define what social history really is (although this is much easier for economic and diplomatic history). In the second place, many historians are still aiming at some form or other of 'total history'. At least they will try at the same time to study various aspects of life such as the economic, the psychological or the cultural side, as well as the interaction of those elements.

Let me now come to my second point: a brief sketch of the history of the historical discipline. For this one has to be a *terrible simplificateur*, and I will be terrible enough to say that we can distinguish two stages or two schools. The first I will label the German historical school of the nineteenth century or the school of historicism; the second I would label the French historical school of the twentieth century or the *Annales* school. This, of course, is only for the sake of argument and of brevity. In fact, the story is much more complicated, but it helps to summarize it in this rather drastic way. We owe very much to the German historical school of the nineteenth century; in fact, we owe it practically everything that was essential for the development of history as a science: all sorts of technical procedures, source publications, seminars, the techniques of text analysis, the use of philology in history, the concept of interpretation, and hermeneutics. In addition, the German historical school stressed the very important techniques of chronology, of how to find exact dates for events which have occurred. We also owe it some important notions that are still part and parcel of historical thinking: the notion of development over time, the diachronic concept of history, as well as the synchronic concept of history, i.e. the notion that every period has its own character and that there is a unity that connects all the phenomena during a certain period. We owe it, furthermore, the notion that goes with it that every period should be judged according to the standards of the time or, in the famous words of the German historian Ranke, that every period is *unmittelbar zu Gott*.

Historicism very strongly underlined the uniqueness of historical events. This led to the great debate at the end of the nineteenth century, a debate (in which Dilthey played a part) that was particularly animated by the neo-Kantians Rickert and Windelband, who claimed that there were, in fact, two models of science, the model of the natural sciences which they labelled nomothetic because it was interested in regularities and thus in laws, and another totally different concept of science, history, which they labelled idiographic. This type of science is interested in the particular and in the unique. Its aim is therefore an accurate description rather than the discovery of laws. This was an important debate but I am not going to insist on that any further.

Historicism and the German school also had some very notable weaknesses. The essential weakness was that it limited itself exclusively to political history. This limitation had many consequences: it means that one studied history exclusively in terms of the nation state, that history was reduced to European history, that European history was again practically reduced to the upper echelons of society, to the study of certain élites. No wonder that chronology was only derived from political systems, the reigns of kings, cabinets, etc.

The second historical school, which I will call the French school, the *Annales* school of the twentieth century, is to a very large extent a reaction to the German historical school. One can perhaps summarize the message of the *Annales* school by saying that it tried to enlarge history. History should be the history not only of Europe but also of other civilizations such as existed in Africa and Asia. The study of the past should not limit itself to the history of élites, but also encompass the history of the masses. This means, as far as the nineteenth and twentieth centuries are concerned, that history should study the working classes, the labourers, and as far as previous centuries are concerned, that history should analyse the position of the peasants, the *laboureurs*. History should not only be political history but also take in economic, social and psychological aspects of the past. Historians should not think only in terms of the nation state, but break this concept down into smaller units such as regions, towns and villages. In turn, historians should attempt to enlarge their studies beyond that of the nation state, and transcend it to larger units such as sub-continents and cultural areas as Braudel has done. History should believe in the chronology of political events but it should also have an eye for the different layers in time and it should use not only the techniques of philology and hermeneutics but also those of the social sciences. In contrast to the notion of the uniqueness of events, the *Annales* school underlined the notion of repetition, of constant elements over time. This development led to another philosophical debate, the debate about history and structuralism in the 1960s, which in turn led to a dramatic rise in the ambitions of the historical profession, this time to develop the study of history not as the other model of science but as the real model of the social sciences. However, there are certain distinctions here. Some people are satisfied when history is labelled as just another social science. Others go a little further by saying that history is really *the* social science, the one and only social science, the ultimate social science. Whatever the outcome of this debate may be, the funny thing is that one cannot say that the new school has replaced the old one. They are actually living very closely together, like the *zwei Seelen* in our breast. In most historians nowadays there is just a little bit of Ranke and a little bit of Braudel and they live comfortably with both.

Let me then come to the third part of my paper, which is the assessment of advancement in the study of history. Of course, I will also follow the distinction between assessments *a priori* and *a posteriori*. This really boils down to the question of which historical books or works historians consider to be the great works of their profession. Put slightly differently, why do historians (and there will be consensus on that) consider some of their colleagues to be great historians and some historical books to be important historical books? I think it is possible to distinguish three categories.

First, there are those historians and those books which give a new vision of a certain period or of a certain problem. For example, when we think of Huizinga – a very famous historian still widely read today – a great deal can be said in explaining why he was such an important historian, but when we come to the discussion of his most important book, *The Waning of the Middle Ages* as it has been translated into English, we see that it all started with a new way of looking at things, a new insight. Huizinga shows in this book that one could look at a certain period – in this case the late medieval world – not so much in the perspective of the dawn of a new era, but in another perspective, actually in that of the waning Middle Ages. We can also think of the historical works of Max Weber and his new vision on the relationship between capitalism and Calvinism. To give another less well-known but still important example, one could refer to the work of the British imperial historians of late nineteenth-century British expansion in Africa, Jack Gallagher and Ronald Robinson. Quite at variance with the traditional vision which regarded the late nineteenth century as the zenith of British imperialism, they presented a new, fresh look at the partition of Africa, and they suggested that the British involvement there should be considered as a decline of Empire, and that the zenith of British imperial history should be found much earlier, around the middle of the nineteenth century. All these publications engendered much discussion among historians interested in such questions. They all offered new insights, a new way of looking at things, and therefore are considered to be seminal books.

In the second category I would place books and works that open a new field of history or introduce a new approach to history. Let me cite as an example the work of the French historian Marc Bloch, who was one of the first to introduce comparative history, and who suggested looking at feudalism not only in a national but in a general context and studying it as a system of authority and power on a world-wide scale. One could also cite the example of Lucien Febvre, who argued that and demonstrated how historians should devote themselves to the study of *mentalité* – the mental attitudes towards several problems – such as death, love, madness, etc. In this category one could also mention the famous historian Lewis Namier, who introduced a new approach to the study of the political system of England in the eighteenth century, understanding it in terms not so much of parties and ideologies but of factions and personal interests. These are all new approaches or ways of introducing new fields within the study of history.

And then finally there is a third category of historical works which are generally considered to be important, although perhaps on a slightly different level. Here I would like to point to books and studies which have enriched the range of historical techniques aimed at coping with sources of a quantitative nature. In fact, it could be argued that quite a number of recently published historical works are based on techniques borrowed as it were from other disciplines such as economics, sociology and anthropology. The French historians Simiand and Labrousse worked out very complicated techniques in order to study the history of prices, while W. W. Rostow tried to construct a model for economic growth in history. New techniques were introduced by Goubert and Henry in developing family reconstruction and historical demography. In addition, attempts have been made at using anthropology in order to write the history of non-Europeans who usually do not possess written sources. Thus, the Amerindian reaction to Spanish expansion in Latin America has been unravelled by Wachtel using oral tradition as a historical source in his book *La vision des vaincus*.

In summing up the impact of historical writing within this third category of new techniques I should mention the study *Time on the Cross* by Fogel and Engerman as an example. In using computerized techniques on quantitative sources on plantation records, as well as applying economic theory on regional economic growth in combination with historical demography, the two American economic historians have turned the traditional historical interpretation of slavery in the southern United States upside down. The findings of Fogel and Engerman have been heavily attacked, but many traditional historians had to admit that the use of quantitative data gave many a new insight which more traditional, descriptive sources were unable to provide.

Perhaps we could refine this scheme, but I think this sort of categorization of historical work into that of new insights, new approaches and new techniques will give you an impression of how historians look at their colleagues in the present, as well as in the past. But, as I said, the *a posteriori* assessment is the easy side of the problem. The *a priori* assessment is, of course, much more difficult.

The only thing I can do is to try to explain to you how research in history is planned in practice and in those few situations where there is some money to give to historical research – which in fact sometimes happens, even in Holland. In the Dutch Historical Foundation, for example, we get some money from the government and have the task of allocating this money to the kind of historical research we consider to be important. I do not think historians will ever come together and say, 'Look, for the next 10 years we have to promote this field of history,' or, 'Now we really have to find out how Julius Caesar was assassinated or whether Dreyfus was really guilty or not guilty.' I have never encountered that sort of a discussion. When historians come together to assess applications for research funds, they ask themselves what sort of people the applicants are, whether they are reliable scholars, have published anything and have research experience, whether it looks as if something interesting will come out of the research proposal and whether the man or woman will ever write the book. In other words, first and foremost the quality of the people who are asking for money will be taken into consideration. That, by and large, will be the most important criterion, however difficult it is to make such a judgement.

In the second place, other things being equal, there might be a certain preference for new approaches. It often happens in a very subtle way: there is a discussion going on in the international scientific world that one reads and hears about; then prospective researchers appear and say they would like to do something on this new type of research, but now, for example, applied to the situation in Holland. The decision-makers will have a certain preference for funding such research proposals because they are new and look promising and could give an opportunity to contribute to the debate in the scientific community. This will continue for some time, but after a while there tends to be a sort of backlash, and a general disinclination to promote this kind of research because it tends to develop into a sort of 'me-too' research. Thus, one tries to avoid the danger of funding an endless series of the same sort of research projects, whose results would be steadily less interesting.

Finally, in the third place, in planning historical research, there might be – in a very vague and unconscious way – some response to what actually goes on in society. In looking back at how things have developed over time, it is possible to note that, at a given moment, more attention than usual was focussed on, say, groups or areas, groups in our civilization, or other areas of civilization, which had been neglected until then. For example, historians may detect that too little has been written on the history of the working classes, or of peasants, or of women, relative to the attention given to those groups in present-day society, and there will be a slight preference to promote research regarding those historical subjects. The same is true for non-European civilizations. African history, for instance, is a rather new field of history, which poses many new problems of methodology. And Africa is obviously a part of the world we should know much more about since this continent has come to the fore in present-day politics. In this way, rather more unconsciously than consciously, society has an influence on what kind of historical research tends to be undertaken.

These three things, then, will be important in decision-making regarding historical research: first, the quality of the people; second, the slight preference towards what are considered new and promising approaches within the field; and finally, the very faint and unconscious way in which people tend to respond to what they feel as a demand from society.

I now come to my conclusion: does all this imply that the field of history should be considered as a science? In some ways, yes; in other ways, no. Returning to the problem of cumulative knowledge as against reconceptualization – as discussed at this colloquium – it is well known that the second element is much more characteristic of historical research than the first. The well-known Dutch historian Pieter Geyl (very well known in the

Anglo-Saxon world) once labelled history as 'an argument without an end', and that is probably the best way to describe what history is about. Historians do ask themselves questions and say to themselves: 'wouldn't it be nice to know'. In this way they might be considered to be scientists, but they are also aware of the many unscientific elements in their work.

The identification of scientific advances in economics

E. MALINVAUD

ALTHOUGH our understanding and mastering of economic phenomena leaves much to be desired and remains quite limited, economics now is a recognized science in which research, accumulation of knowledge or fundamental reconsiderations occur, as they do in other sciences.

But two reasons make the identification of its scientific advances particularly difficult. In the first place, there are far too many attempts at bringing the general public up to date on the supposed evolution of economics; a new idea or trend of thinking, if simple enough, is often advertised before being properly discussed and tested. This state of affairs can be understood, considering how economic conditions react on everybody's life; but it means that non-economists may often think that a new 'advance' has been identified by economists. Once they are aware of this practice, common sense and experience advise them to take any information about economics with a good deal of scepticism.

In the second place, economists themselves are divided. There is no universal agreement among them on what is the present state of scientific understanding of economic phenomena. Of course, one should not overstress this point: division among scientists exists in all fields of knowledge; moreover, in many organizations dealing with applied economics, people adhering to different paradigms work peacefully together, which shows a much higher degree of scientific agreement than the layman would expect. The division is, however, serious enough to make the internal identification process somewhat difficult. One may say that a commonly recognized process does not exist in economics by means of which scientific progress could be asserted. In a fair proportion of cases, some developments that are qualified as being advances by certain economists will be considered as useless or negligible by others.

The criteria applied in such evaluations can be perceived on two levels. On the one hand, one may say that economists subscribe to the general principle according to which scientific progress means improvement in the ability to predict and prescribe, a principle that is quite congenial to the object of most studies of applied economics; but whether there is any direct or indirect improvement in this ability often remains debatable. On the other hand, one may look at the criteria applied in the day-to-day evaluation of academic research contributions; here again, the criteria will not look surprising, but every school of thought would apply them differently, if they were all actually confronted with the same cases. (Most often, each school evaluates the contributions of its members with little concern for what is happening elsewhere.)

To give a proper account of this complex situation, to show how some advances are recognized, either internally or externally, and to exhibit correctly the importance and role of divisions, may not be easy. One may, however, organize the subject, using the distinction between analytical research concerning narrow and well-defined topics, and the construction of general systems explaining broad categories of phenomena. Such will be the focus of attention in the first and second parts of this paper. The third part, devoted to a case study of Keynesian theory, will be intended to complete the picture and to raise problems arising between internal evaluation within the community of scientific economists and the recognition of advances by the learned public.

1 INTERNAL EVALUATION OF SPECIFIC RESEARCH WORK

In economics, scientists working on any of a large number of specific topics usually form a group within which criteria of scientific achievement can easily be found. For many such topics, no fundamental conflict would exist between the evaluation given by various schools.

In order to convey a concrete feeling about the prevailing situations, examination of the criteria applied will proceed by a brief survey of a selection of topics, this selection being intended to represent the whole spectrum of analytical research in economics. One will then consider what can be the global outcome of this evaluation process, operating at the level of many specific research projects.

1.1 Measurement of economic evolution

The pure observation of economic phenomena, at some level of generality, deserves good scientific work. The definition of the concepts to be measured is neither obvious nor unimportant. Comparative assessments between periods, countries and social groups raise a number of problems concerning both the collection of the basic data and their treatment. This is why research efforts have been devoted to this subject for a long time.

The most important advances in this field have combined methodological aspects with more or less extensive data collection. The choice of the basic concepts is hardly ever easy. To take just one example, the definition of 'income' required serious consideration by many economists – among the best known I. Fisher (1911) and J. Hicks (1939) – but it remained imprecise before it was codified in the internationally agreed systems of national accounts. Even today, the definition incorporated in these systems is not quite appropriate for all purposes and some research workers have to revise it in order to give better measurements of the phenomena they are considering.

The profession of economists of course attempts to be ever more rigorous in its use of basic concepts. This is, however, achieved in a diffuse manner, so that it would not make much sense to attempt to analyse this evolution in order to identify the origins of major changes. The language of economists, i.e. the basic set of concepts they are using, is now fairly well unified; if Marxists' abstract writings still use some vocabulary of their own, they have no special problem in referring to the general stock of available statistics and economic accounts when they want to turn to applications.

On the methodological side, questions also concern the treatment of the basic data in order to derive meaningful aggregate measurements from them: how should one compute an index of prices, or one of output, how does one characterize inequality in incomes at a particular date and its changes through time? Here, the contributions of individual scientists are easily identified. They may be said to be significant if they are actually used by others, either in their empirical or in their methodological work.

But, considering how scanty the factual assessment of important aspects of economic phenomena still is, collection of data is often more valuable than methodological refining. Here again, one will have no surprise in learning about the criteria used to evaluate particular contributions: the most praised will be those bringing new results, covering important historical segments, and succeeding in achieving the best feasible accuracy. Scientific disputes of course occur on occasion on what are or what were the facts. But they do not seem to reveal any feature that, particular to economics, ought to be reported here.

1.2 Econometric finding of agents' behaviour

The behaviour of individuals, households, firms and other organizations has to be precisely known if economic phenomena are to be understood. To some extent, behaviour is revealed by theoretical reflection about the conditions under which, and the purpose for which, each one has to act. But theory alone does not lead to a full characterization. Reference to observed data is also necessary in order to infer from them the precise role of each one of the many factors that are *a priori* likely to affect behaviour. This is why econometric research is invaluable for the progress of economics and is indeed very active.

A good econometric study requires good data, a good model and a good statistical estimation procedure. This is clear. In particular, the model must not prejudge what is still unknown and what the data are supposed to reveal; on the other hand, it must translate, as precisely as possible, the solid conclusions derived from prior theoretical analysis or from prior econometric analysis of other data. Specification of the model is therefore a delicate operation.

When a new econometric study, or group of studies, is becoming available, concerning particular behaviour (for instance, investment behaviour of industrial firms), its evaluation proceeds through consideration of such questions as the following. Is the specification better founded or more general than preceding ones? Is a new, more accurate or broader data base being used? Is the statistical inference more powerful or more robust against specification errors?

There is not much to say that would be of general value and would distinguish the inference process in economics from that occurring in other sciences, except that controlled experiments are impossible. Hence, the most efficient way of testing scientific hypotheses is not available to economists. They must rely on repetition and confrontation of econometric studies. They often have difficulty in discriminating between two theories that happen to account about as well as each other for the stock of observed facts. Discrimination does occur, or will eventually occur, but as a result of a very slow evolution, during which protagonists will often find apparently good reasons for explaining why their own theory did not work well in some particular cases, which, however, do not disprove the theory.

1.3 Structures, performances and valorization of capital in industry

The most challenging field for modern economists concerns the internal operation of industry,

> with its complex system of firms, having very different sizes and structures, from the large multinational corporations to the thousands of small undertakings;
> with its variety of 'markets', some concerning a well-defined commodity with no direct substitute, others involving supplies of many somewhat substitutable manufactured products, some served by monopolies, others by a large number of competing firms;
> with its heavy installations, its specialized personnel, its long-term contracts, its complex connections with public authorities.

In view of these complexities, one should not be surprised to find that the scientific modelling of the economic operations of industry is very little advanced, notwithstanding its obvious relevance. Many economists, however, work for it, some coming from the neo-classical tradition, others inspired by the Marxist vision. Fundamentally, their methodology is the same, namely to develop an appropriate system of concepts for detailed investigation of these economic structures and operations, and then to proceed to measurement and hypothesis-testing on such extended geographical and historical areas that the results obtained may have some general validity.

Criteria for the identification of scientific advances directly follow from this kind of approach. One wonders whether some interesting new concept or measure has been found to characterize concentration, specialization, competition, barriers to entry, profitability, valorization of capital, labour force management, workers' activity, etc. One asks whether the factual conclusions derived from new studies significantly add to pre-existing knowledge or lead to revising it substantially.

1.4 Mathematical economic theory

A good deal can be learned about economic phenomena by purely abstract reasoning starting from non-debatable hypotheses. Indeed, we have direct knowledge of the objects and institutions involved; we know much about the conditions under which agents behave and also about the objectives that they pursue. Hence, abstract theory has had an essential role to play, in order to discover the outcome of the simultaneous behaviour of agents that are quite interdependent. Naturally, the application of greater rigour in theory has made it more and more mathematical.

Mathematical economics concerns a great variety of subjects and should not be identified with one particular theory. It deals with the way in which the allocation of resources can best achieve its aim, with the determination of prices and wages in competitive or capitalist economies, with the process of economic growth, with unemployment and inflation, with international trade and the international transmission of business conditions, with industrial structures and markets organization, and so on.

Criteria of mathematical rigour have become more strict than they used to be; but they still vary substantially from one branch of economic theory to another. When a mathematical system is built in order to represent a theory and to see how some variables react to changes brought to exogenous conditions, it is not always the case that existence of a valid solution to this system is fully discussed. Hence, the assumptions necessary for the completeness of a theory are not always precisely known. Implicitly, economists accept a division of labour between those who first specify or explore a theory (innovative research) and those who will consolidate it at a later stage, making it more rigorous.

The incomplete rigour of mathematical economics explains why debates concerning the validity of some purely abstract propositions may have occurred, such as that between the two Cambridges about the possibility of defining a perfect aggregate for the volume of capital. The passions that surrounded the discussion explain why it was so confused. Eventually, the point was, however, settled.

The most delicate question raised by the evaluation of new contributions to mathematical economics concerns their relevance. Editors of journals often wonder whether they should accept mathematically correct articles whose significance for economic theory is difficult to assess. Two referees will sometimes give opposing advice in this respect, even though they will both be active in the same area of research as the one explored in the paper and will, as a rule, tend to value this area highly, often more highly than is justified.

It is, of course, about the relevance of economic theories that people belonging to different schools will most often disagree (see J. Pajestka and C. H. Feinstein, 1980). The disagreement may sometimes concern the object of the theory; for instance, proponents of *laissez faire* may not be interested in the theory of planning. More often the disagreement will, however, bear on the positive representation that is given of economic reality: how far can one go in explaining economic phenomena while sticking to the hypothesis of perfect competition? Is market clearing or disequilibrium between demand and supply the dominant feature of the situation? Such questions should eventually be answered by rigorous tests from the facts; but, as was explained above, for a less difficult kind of study, conclusive tests are not easy to find.

1.5 Global outcome

What is the end result of the research evaluation that currently takes place in many fields of economics? In order to answer this question, one should, of course, not pay attention mainly to the few research topics that happen to be fashionable now, but rather sit back and take a broad perspective. An economist writing for readers working in other sciences may not be in the best position to do so and should also avoid overselling his discipline.

I shall therefore leave the question open, except for two comments. On the one hand, as the preceding survey of a number of fields suggests, progressive accumulation of knowledge is the typical pattern of research in economics. Of course, this cumulative process is stimulated by the realization that some important questions remain unsolved and the agenda of questions to be solved changes through time, but most often it does so slowly and somewhat unconsciously. The following pages will show that breakthroughs and innovations have also occurred in economics, but real ones have been rare, much more so than the information reaching the general public through the media would suggest. On the other hand, the subject of economics is progressively becoming more rigorous. Factual or logical propositions can be falsified, and have been on many occasions. A few examples may help to show how.

A frequent view in the middle of this century was that economic growth would automatically reduce social inequalities and greatly benefit the poor. The careful study of actual evolution in many countries of the world has shown that such a thesis was much too simple, that changes of inequality tend to be typically slow and that they are perverse in many cases, notably during the initial phases of fast economic growth.

If mathematical economics has developed many general theories that provide invaluable references for a serious reflection about many phenomena, it has also found that few general propositions hold without more restrictive assumptions than was first thought. Demand curves are not always downward-sloping, the accumulation of capital does not always depress the rate of profit, discounting does not always make the distant future unimportant, and so on. The extra hypotheses that are sufficient to save the preceding propositions are interesting to know in each case.

Unqualified statements about the efficiency (or inefficiency) of capitalism have been disproved in many ways. Conditions for efficiency turn out to be very complex, once one takes proper account of many aspects of economic activity, such as time, uncertainty, the actual distribution of information, irreversibility, the collective nature of many services, and so on. The scientific achievements of the past few decades have completely removed from economics (but of course not from the stand taken by some economists) its vindicative nature in support of any particular form of society.

Taking account of the latter evolution of economics does not mean that broad theoretical systems have disappeared, but rather that they no longer claim to provide perfect representations. Each one of them should now be viewed as organizing the thoughts and research about a large range of phenomena, considered from one important point of view, as suggesting new investigations, as incorporating their results within a coherent general model, and as providing a framework for a discussion of the likely impact of economic policies or institutional changes. Economists can often use with benefit two or several of these systems in order to grasp complementary aspects of the same phenomena.

2 BROAD THEORETICAL SYSTEMS

The role of such systems and the manner in which advances concerning them are identified deserve special attention. All the more so because, in a number of cases, economists disagree as to the usefulness and significance of some of these systems.

In order to convey a correct feeling for the situation that prevails in this respect, the best approach seems again to be the consideration of a number of cases, which should be chosen in such a way as to be somewhat representative of the existing diversity. When selecting four such broad systems I have tried to consider important ones and refrained from discussing recent developments, which will perhaps appear negligible in a few years. I have also decided against reporting on the theory of the general competitive equilibrium, because it would have required too much space and might thus have reduced the attention given here to other general systems. The next part will be concerned with the fuller examination of one case and might have been devoted to the competitive equilibrium, but the Keynesian theory is more interesting from the point of view of this conference.

The systems to be discussed may be said to play the role of paradigms and, for two of them at least, cannot yet be considered independent from prior ideological stands. At present hardly any economist will therefore dare to pretend neutrality when considering them and I shall not do so. The reader will, I am sure, have this fact well in mind; he would not have been correctly served if I had decided against writing about somewhat ideologically loaded theories, since they are, after all, parts of the picture.

2.1 The theory of games

The relevance of imperfect competition has always been recognized in economic theory. Actually, the first models of imperfect competition appeared in the middle of the nineteenth century, at the time when the concept of perfect competition really emerged. But the theory proceeded by the independent examination of a number of cases that could not pretend to cover a representative sample of actual situations: monopoly, bilateral monopoly, monopolistic competition and so on. Even the apparently simple duopoly case gave rise to many distinct subcases.

When, in 1944, J. von Neuman and O. Morgenstern published their *Theory of Games and Economic Behavior*, it was soon declared to be a major innovation, a real breakthrough. The book proposed a general approach to the study of decisions simultaneously taken by agents whose interest may be more or less conflicting but also more or less concordant; moreover, it accounted for the occurrence of chance events that sometimes play an obvious role in economic life. Economists were flattered by the fact that one of the best mathematicians of his time had turned his attention to their subject. More fundamentally they accepted the claim of the book, namely that it provided in particular a unified and efficient approach for the theoretical treatment of all situations of imperfect competition. They then entertained the expectation of seeing the quick development of a powerful theory that would make economics more realistic.

But the progress turned out to be very slow. Still worse, it appeared that the basic concepts for games with more than two players were subject to what appeared to be considerable indeterminacy. Thus, none of the previously examined special cases of imperfect competition received a new and unambiguously accepted treatment. It is probably fair to say that, at the beginning of the 1960s, most economists had fallen into complete disenchantment and many of them tended to consider that the idea of applying the theory of games to economics was a blind alley.

One realizes today that the latter pessimistic view was also wrong. Research into the application of games theory to economic questions was very active, and on the whole rewarding, during the last 20 years. On the one hand, we progressively learned how to use efficiently the various basic concepts of this theory. On the other hand, we applied it to other relevant economic questions beyond the precise modelling of imperfect competition; for instance: what are the conditions for perfect competition; how should one choose the incentives of individual decision-makers in decentralized economic organizations?

The theory of games did not bring the revolution that was once expected by some of us, but it now definitely contributes to the progress of our science; we could no longer live without it. It was truly a breakthrough, opening new possibilities for the cumulative and laborious process of relevant theory construction.

Is there now a consensus among economists about the mature evaluation that I just gave? I tend to believe it, but I have no strong proof. The subject does not actually rouse controversies that would reveal the thoughts of various parties.

2.2 Monopoly capitalism

In his statement of some laws of human history, Karl Marx insisted on the major role played by the social transformations following from changes of the techniques and organization of production. The historical succession of 'modes of production' is moreover, according to him, strongly determined. From these premises a theory of capitalist development results. 'Monopoly capitalism' is a short designation of the stage reached by this development in the twentieth century.

A number of economists take this theory as the main framework for their research. They want to characterize the trends of the structural changes occurring in modern economies and to relate them to social transformations. They give special attention to the analysis of profitability of various types of enterprise or nation, in relation to the way in which they are managed (what is often called 'the valorization of capital'). They try to incorporate in the picture working conditions, labour relations and political conflicts. They emphasize the various malfunctionings that are likely to give rise to necessary changes (the 'contradictions' of capitalism).

A question in this conference is to know how scientific advances in this theory are assessed. The question should not be confused with the evaluation of particular results obtained by economists working in the above framework, such an evaluation proceeding along lines that are very similar to those applied, for instance, for research on the measurement of economic evolution or on industrial structures (see sections 1.1 and 1.3). The question concerns the broad system that the Marxist theory of capitalist development proposes.

A difficulty for assessment in this case comes from the debatable nature of what should be considered as scientific advance. The theory aims at providing a very embracing vision of development and is not precisely formalized. It is not often clear whether new statements of this vision are intended to modify it, to make it more precise or simply to restate it and stress its value. No formal test of the proposed changes is provided. Some might perhaps even say that the notion of scientific advance should not be used here.

Economists do, however, on occasion assert views about the scientific progress of this theory; but then agreement does not seem to occur. For instance, the proposition according to which the underdevelopment of the Third World is a consequence of the development of western capitalist countries, even independently of imperialistic domination, appeared, I believe, in a clear form in the 1950s. Many Marxist economists, if not most of them, will identify the emergence of this proposition as scientific progress. But many other economists will say that the proposition was not proved, even among those economists that are ready to use the notion of conflict of interests between the industrial 'centre' and the Third World 'periphery'. Some economists will even say that the proposition is absolutely wrong.

2.3 General equilibrium with sticky prices

The notion that many economic phenomena are related to some more or less durable imbalance of the price system is very old in economics. The study of economic evolution has always stressed the importance of intersectoral or geographic differences between the remuneration rates earned by identical factors of production; these differences explain factor movements and other dynamic phenomena. On the other hand, reflection on the foundations of the Keynesian theory of unemployment has shown that it was built from two essential notions, that of a discrepancy between supply and demand, at least in the labour market, and that of a general interdependence of the decisions simultaneously taken by various economic agents. Discrepancies between supplies and demands are related to the fact that some prices are too sticky to clear the markets to which they apply.

Ten years ago, the full formalization of a temporary general equilibrium with fixed prices was worked out, incorporating the two essential notions used in Keynesian theory and proposing a rigorous building block for the modelling of some dynamic phenomena of economic evolution. As for the logical soundness of this contribution, there is no disagreement. It is recognized as being a good piece of mathematical economics, even though it raises some puzzling and as yet unsolved questions. But the significance of this contribution is seen quite differently by various groups of economists.

Some economists, mostly western European ones, consider it as a very important innovation that will permit and make more rigorous a large part of economic theory, that integrates into a consistent system previously distant branches of economics, that exhibit the nature of, and reason for, some phenomena attracting the attention of practitioners but previously neglected by theoreticians. Other economists, mostly working in American universities, have taken a strong stand against the significance of this new development. They argue that it relies on *ad hoc* theories, in particular that the non-clearing hypothesis on which it is based has no proper foundation and contradicts the basic law of economics, namely that no mutually advantageous trade can remain ignored by economic agents.

This conflict will probably soften in the years to come, since the matter is rather recent and the positions argued by strong personalities at first obtained adhesion from people who may not have thought sufficiently about the issue. But it is revealing of the difficulties faced by assessment of scientific advances concerning broad theoretical systems.

Part of the problem undoubtedly comes from what is at stake, namely an evaluation of the need for government interventions on the economy. If market clearing always occurs, the case is weak; if, on the contrary, undesirable sustained imbalances exist, most people will think that something should be done about them. This background explains why scientific evaluation is indeed polluted by ideological considerations, in this case as in the preceding one and in the following one.

2.4 Monetarism

The same background lies behind positions taken about modern monetarism, a thesis on economic phenomena that is not easily identified, except by reference to its main protagonist, Milton Friedman. Briefly stated, the thesis amounts to belief in propositions such as the following:

money is important;
short-term economic fluctuations cannot be stabilized;
monetary policy should aim at a constant and properly chosen rate of increase of the aggregate quantity of money.

It is not the place here to discuss such propositions, nor to evaluate the analytical or econometric contributions due to Milton Friedman, many of which are commonly recognized as having been important, nor either to describe how modern monetarism was developed as an alternative to Keynesianism, an aspect that will be touched on in the next part of the chapter. It would be possible to consider the various steps of this development here and to examine whether and how they have been identified as scientific achievements; but I shall concentrate my attention on just one of these steps, which brings into the picture a different kind of economic research from those discussed in the preceding pages. I shall call it 'the empirical proofs of monetarism'.

The starting-point was the observation that time series of the two main economic aggregates, the volume of production and the general price level, can be fairly well fitted to time series of past and present values of the quantity of money, at least in some important countries. According to the relationships so obtained, an increase in the quantity of money would be followed first by a temporary increase of output, and later by a permanent increase of the price level.

In order to interpret these observed relationships as proofs of monetarism, one must not only check that they are held in all countries, but also decide whether they should be given a causal meaning and, if so, whether other instruments of economic policy do not play a role, besides the quantity of money, on a par with it, or even before it. More or less complex inductive procedures have been devised to serve for this decision. The discussion on these empirical proofs started around 1960 and is still continuing, becoming more and more sophisticated.

Two remarks are worth making in this conference. First, the discussion was undoubtedly considered important by the few well-known economists who often sit on the stages where debates about macroeconomic theory are given. It also attracted a small number of other economists, but it has remained so far definitely less popular than other less empirically oriented issues, such as the consequences of assumed 'rational expectations'. This may reveal some more general difficulties that have always faced attempts to give a strong empirical basis to economics.

Secondly, in the course of this discussion the force of the existing proofs has been evaluated quite differently by the two sides. Proponents of monetarism argued that those were indeed definite proofs, while their opponents declared themselves unconvinced for various reasons. One side was prompt to identify a scientific advance, the other asked for more conclusive evidence. Again, one cannot help thinking that this difference of attitude has much to do with the policy implications drawn from monetarism.

3 A CASE STUDY OF EXTERNAL EVALUATION: KEYNESIAN THEORY

Up to this point we have been concerned with the identification of scientific achievements within the profession of economists. But something must also be said about external evaluation of economic theories by the general public. This can best be done if we select one example; Keynesian theory, then, is the obvious candidate for a particular examination.

By Keynesian theory I mean what was taught almost everywhere throughout the world in the 1960s about the determination of short-term economic evolution. This theory owes much, of course, to *The General Theory of Employment, Interest and Money*, published in 1936 by J. M. Keynes. It should, however, be distinguished from it, not only because it is free of the ambiguities of that book: it simplifies and ignores a number of points made by Keynes; it also supplements it by the juxtaposition of a dynamic process that has been called 'the Phillips curve'.

For the learned public this theory slowly emerged in the 1940s and 1950s while it was still considered somewhat debatable, it was recognized in the 1960s as providing the reliable framework thanks to which economic policy could be correctly chosen, but the theory fell into disrepute in the 1970s. Such an account is not a caricature, although it applies only to the dominant view from which deviations of course existed. *

* It would hardly be a caricature to push further and say that in the 1940s and 1950s, economics slowly emerged as a respectable subject; in the 1960s it was recognized as a science; but it fell into disrepute in the 1970s.

Considering how this change of mind occurred and developed suggests three conclusions that we shall examine in turn:

the public is somewhat fooled by the media and their proneness to oversimplification;
the public confuses good (bad) economic conditions, or apparently efficient (powerless) economic policies, with sound (misleading) economic theory;
external evaluation in turn reacts on research evaluation and research programmes within the profession.

3.1 Oversimplification by the media

It must be granted that the change of mind of the general public reflects, but unduly magnifies, an evolution of ideas that took place among economists. Many of them were slow to understand the new conceptual system that was proposed by Keynes and conflicted with the full market clearing hypothesis on which the previous progress of economic theory had been based. Later, some limitations of Keynesian theory were overlooked or said to be unimportant. We have now realized that they should be seriously examined and that taking them into account may lead to substantial revision of the analysis and of its conclusions; Keynesian theory is, now more often than before, regarded by economists as incomplete (but not as useless).

How oversimplification is misleading can best be understood on the particular issue of the Phillips curve. When in 1958 A. Phillips published his article showing a strong inverse correlation in the United Kingdom from 1861 to 1957 between the rate of increase of nominal wages and unemployment, the phenomenon did not surprise economists since it looked so much like the celebrated law of supply and demand. The article certainly attracted interest, was even identified as a major contribution to macroeconomics and definitely influenced subsequent thinking and research about it. But the article was not qualified as being the breakthrough that the media later described. The surprise was rather that the phenomenon would appear so neatly that unemployment alone could explain the speed of nominal wages. But one soon realized that indeed other factors should also with benefit be taken into account; this was pointed out in an article by R. Lipsey that appeared hardly more than a year after that of Phillips.

This inverse correlation poses a dilemma to macroeconomic policy. Policy-makers often have the two simultaneous objectives of raising employment and reducing inflation, of which wage inflation is a part. Hence, they must be conscious of a trade-off between unemployment and inflation. Other things being equal, the trade-off may be graphed as a downward-sloping curve, the Phillips curve. But other things are not always equal; depending on what they are, the curve may at any time be higher or lower, and the dilemma facing macroeconomic policy more or less challenging.

This explains why econometricians, stimulated by the initial work of A. Phillips, did not aim at determining a fixed curve, but rather at finding out a wage equation in which the slack on the labour market would be one among other explaining factors. They recognized that the role of this factor may vary depending on 'institutional' conditions; but they found in many cases that the wage equations fitted on pre-1974 time series were still appropriate after the first oil shock.

The view that was given to the general public by the media was quite different: in the 1960s the fixed curve notion of the trade-off was popularized; in the 1970s it was reported over and over again that 'the Phillips curve had failed'.

3.2 Confusion between the state of the economy and the state of economics

People who have been cured are more likely than others to think medicine is a science. At the time of the fast expansion of the 1960s it was easy to entertain the idea that economics had progressed so much that depressions would never again occur, and this thanks to Keynesian theory. Considering the present economic conditions, it is tempting to conclude that any existing theory about economic stabilization is wrong, since governments are apparently unable satisfactorily to stabilize the economy.

One should not, of course, dismiss confrontation between economic events and what economics say about them. Indeed, this confrontation is healthy: it plays an important role in the internal evaluation of research efforts and in the change of ideas within the profession. But whereas the conclusions drawn by the general public about Keynesian theory were pretty much those suggested in the preceding paragraph, most macroeconomists drew a different conclusion, namely that Keynesian theory paid too little attention to medium-term effects of economic policies.

Indeed, this theory, supplemented by econometric models based on it, quite correctly predicts what will be next year the consequences of stimulative or restrictive actions taken today. But it does not pay attention to what may result after 10 years of repeated use of policies inspired by short-term considerations. Even if each year the correct diagnosis has been made for next year's situation and correct measures to improve this situation have been taken (history shows that this assumption is not always fulfilled), the policy may unfavourably react on the situations that will have to be faced in subsequent years. Moreover, it appears that Keynesian theory, with the Phillips curve incorporated in it, does not deal well with these medium-term consequences; it must be supplemented, or perhaps even replaced, by other theoretical constructions when attention focusses on such consequences. Keynesian theory should not be discarded, but its limitations should be known and alternative theories ought to be found for dealing with issues that have now become important.

3.3 Impact of external evaluation

Economists would have good reasons for remaining insensitive to judgements formed by the general public about their science. After a period of excessive confidence and another period of excessive disparagement, something else will undoubtedly come. Indeed, most scientists among us go on undisturbed, pushing ahead their own research programme, paying attention only to judgements given by peers. Some writers, of course, seize the opportunities offered by the market and produce supposedly inspired books that comfort public opinion and try to explain how economics ought to reform itself or the economy ought to be reformed, but the impact of these books within the profession is negligible.

Reaction of economic research and economic scientific thinking to external evaluation is therefore limited. It, however, occurs whenever this evaluation reinforces or weakens the position of one side in a scientific debate. For instance, the external dismissal of Keynesian theory increased the interest given by the profession to monetarism, which claimed to provide an alternative theory.

First, the arguments raised by monetarism had to be taken more seriously as soon as they were echoed in the media. Non-believers had to sort out these arguments, to clear up their own thinking on related points, to look again at the data for an answer to new questions. As always in a dispute, they had to give up some of their previous claims. Secondly, research along monetarist lines was given a new impetus. The allocation of research funds by government agencies or foundations could not remain insensitive to the prevailing interest for alternatives to Keynesianism. Young people were attracted by what was said around them to be a promising development.

4 CONCLUDING REMARKS

This survey of research evaluation in economics probably reveals a good deal of similarity with what could be said about other sciences. Notions that are common elsewhere appear also when the scientific evolution of economics is considered. One can speak of the 'frontiers of knowledge' about investment behaviour, of the 'breakthrough' that *The General Theory* provided, of the 'integrative advance' offered by the fixed-price general equilibrium, of the 'innovation' that the first statement of the Phillips curve was, of the 'decline' of Keynesian theory. Since economics can claim only a limited understanding of economic phenomena, important unsolved problems exist in almost all of its branches. Moreover, assessment criteria, which are not dissimilar to those used elsewhere, are probably applied with less stringency.

There are 'accumulative imbalances' in economics also. The refinement of the mathematical theory of perfect competition, as opposed to the embryonic state of the theory of other forms of competition, is a well-known and well-recognized case, considering the facts of life. Whether there is an imbalance between theoretical advances and accumulation of data, the latter being lagging, has often been discussed and is still a pertinent question to raise; the positive answer, that most economists give, remains, however, somewhat disputable, considering the strong objective obstacles that limit the power of inductive research in economics.

Can scientific advancement in economics be systematically oriented by a conscious research policy? My own answer is quite cautious. On the one hand, the achievements of econonic research during the past 30 years owe much to the place that was given to it in teaching and in the allocation of research funds; progress in our econometric knowledge would, in particular, have been impossible without the costly collection of data that took place. On the other hand, it seems to me that the distribution of resources among economic research areas hardly ever achieved what was intended; fortunately, it was often diverted to the benefit of spontaneous and unpredictable imaginative efforts in related areas.

What may be special for economics does not come only from the rather early stage of its present development but also from its proximity to political issues. Attempts at progressing in the knowledge of economic phenomena are not always detached from direct involvement in policy formation. The requirement of scientific rigour then conflicts with the wish to influence decisions quickly. Economists ought to be conscious of the risk that such a situation implies for the advancement of their science. Other scientists, when turning their attention to economics, must also be aware of this risk.

Comment

L. PASINETTI

As I agree with much of what Professor Malinvaud has said, I would like to approach the subject from a different angle. To begin with, we may ask what the particular features of economic research are. More specifically, what are the peculiarities of economics, as compared with physics, considered by many – as Professor Haken has said – as 'the prototype of science'?

I shall start by listing four of these peculiarities. First of all, unlike physics, astronomy, medicine or many other subjects, the *object* of economic studies is changing continually. When Ptolemy, 22 centuries ago, observed the planets and stars in the sky, he was looking at exactly the same universe which we explore with our own radio-telescopes today. The universe has not changed (or has changed in an absolutely negligible way). We simply understand it better today. We are able to penetrate it further and deeper. But this is not the case in economics. When Adam Smith, only two centuries ago, was enquiring into the English society that was emerging from the Industrial Revolution, he was looking at something, not negligibly, but quite profoundly, different from, let us say, Thatcher's England (or, for that matter, Mitterand's France) in 1983. Not only are economists unable to reproduce the phenomena they are interested in, as Malinvaud has pointed out; sometimes it may not even be useful to reproduce them. The economists must clearly aim at explaining the events of today. For that purpose, those of yesterday may or may not be of great assistance. In any case, they are the only ones that can be observed.

Secondly, economic research (again unlike physics) will tend sometimes to influence the events that are the object of study. The economists cannot simply stand back and observe, or explain, in a detached way. When they try, for example, to understand the causes of unemployment or of inflation, they obviously do so with the aim of devising methods to remedy such undesirable phenomena.

Thirdly, economics deals with some simple, down-to-earth, day-to-day problems that are everybody's immediate concern. It is enough to think of devaluation of the currency, of the rising level of prices, of the movements of wages and salaries, to realize how deeply these phenomena affect everybody's pocket. This explains why ideas, hypotheses, even guesses, are avidly seized by the mass media, sensationalized and very quickly put across to the general public, even when they are still at a tentative stage. Malinvaud points this out very well.

Fourthly, many pronouncements on economic matters cannot avoid value judgements. Think of measures affecting the distribution of income or of wealth, or concerning the tax system, the level of prices and interest rates, etc., let alone the value judgements at the basis of the 'vision of the world' which inevitably lies behind even abstract theorizing.

Yet, after listing these characteristic peculiarities of economic research, it must also be stated that they do not cover the whole field. More specifically, not everything that is studied by economists is continually changing. There are important features of the economic system that may persist for quite long periods. Some implications of the process of division and specialization of labour, for example, or some basic features of the pattern of evolution of consumers' spending (the so-called Engel curves), remain as relevant today as they were at the time of Adam Smith. Again, not everything that the economists investigate will, when understood, be used to influence events: far from it. Sometimes the economists turn out to be powerless even when they think they should be able to influence events. Again, not all subjects of economic analysis are of immediate concern to everyone. Some abstract elaborations in game theory, in intertemporal optimization theory, etc., are as far away from everyday life as the most abstract of philosophical speculations. And finally, not every statement the economists may make, not every conclusion they may reach, is conditional upon value judgements. What could be more objective, for example, than an econometric estimation of demand elasticities? This extreme heterogeneity of the characteristics of economic research, when compared with other disciplines, might itself be listed as a fifth peculiarity.

It emerges quite clearly from these remarks that these peculiarities are bound to create difficulties in communication between economists and the public at large and among economists themselves, not least because, as Malinvaud points out, the public at large tends to confuse economic theories with economic events. But how do the economists themselves react to these difficulties? How do they consider their theories? As examples, I have selected a few statements by two distinguished economists, belonging to different schools: John Hicks, a Nobel prize laureate, from Oxford, England, and Richard Goodwin, an American economist originally from Harvard University, transplanted to Cambridge, England.

In a methodological essay in 1976, Hicks made the following claim:

> In order that we should be able to say useful things about what is happening, before it is too late, we must select, even select quite violently. We must concentrate our attention . . . We must work . . . in some sort of blinkers. Our theories, regarded as tools of analysis, are blinkers in this sense. Or it may be politer to say that they are rays of light, which illuminate a part of the target, leaving the rest in the dark . . . But it is obvious that a theory which is to perform this function satisfactorily must be well chosen; otherwise it will illumine the wrong things.

But who is going to decide which are the right things and which are the wrong things to illuminate? In 1983, Goodwin was more specific:

> In my view, economies are so impossibly complex as to defy any completely satisfactory analysis: rather the best that can be hoped for is a number of different approaches, each of which yields valuable, but incomplete insights into the various aspects of the system. Thus 'general equilibrium' theory, whilst in principle admirable, always has seemed to

me to be so 'general' as to be largely vacuous and even capable of diverting attention from important matters. By contrast Keynesian theory and cycle theory, for all their crudity, attracted me because of the usable practical results.

Implicitly, from the methodological approach taken by Hicks, and quite explicitly from Goodwin's attitude, we can see here the emergence of *paradigms*. I was surprised at Malinvaud's not mentioning them, except in very incidental passages or implicitly when he talks of 'different schools', but it is quite clear that it is only by allowing the existence, or more precisely the coexistence, of different paradigms in economics that one can grasp the full meaning of various statements in Malinvaud's paper, such as the following:

. . . a commonly recognized process does not exist in economics by means of which scientific progress could be asserted.
[Economists] often have difficulty in discriminating between . . . theories . . . Discrimination . . . will eventually occur, but as a result of a very slow evolution, during which protagonists will often find apparently good reasons for explaining why their own theory did not work well in some particular cases, which, however, does not disprove the theory.

All this leads me to end my comments by stressing remarks and recommendations very similar to those that have been made for other fields of research. For economics, also, I think one can repeat: beware of citation clubs, and be conscious of the responsibility of committees awarding prestigious prizes, which inevitably generate bandwagon effects. But most of all I think it is important to stress the heavy responsibilities of the committees that are in charge of the allocation of research funds, not only because of the difficulty of 'predicting the unpredictable', as was said by Professor Haken, but also because, to take another remark of Professor Malinvaud as typical of research on economic subjects,

. . . it seems to me that the distribution of resources among economic research areas hardly ever achieved what was intended; fortunately, it was often diverted to the benefit of spontaneous and unpredictable imaginative efforts in related areas.

This being the case, it is important that, also in the allocation of research funds, one should take into account the coexistence in economics of different and competing paradigms, to be considered, as said by Professor Dik, as 'signs of vitality rather than immaturity' with 'discussions across paradigm boundaries to be stimulated rather than shunted'. Perhaps, just because so much is at stake (in terms of value judgements and ideological positions) the economists may not have been particularly far-sighted on this matter in the past. But this is a further reason for drawing attention to the problem of the responsibilities of committees in charge of the allocation of research funds.

Breakthroughs in ecology

L. B. SLOBODKIN

1 INTRODUCTION: THE CURIOUS ROLE OF ECOLOGY

THERE is an important distinction between the science called ecology, which is an academic subdiscipline originating in biology, and the various aspects of what is called the ecology movement. Partisans of the ecology movement typically advocate a normative standard of political actions which are more or less tightly tied to an ecological life-style. The ecology movement may have grown from, or been strengthened by, the results and popularization of the results of ecology; however, most ecological research workers are not members of, partisans of, or even strongly cognizant of, the environmental movements. I shall be concerned with the scientific discipline of ecology, not the ecology movement. The notion of breakthrough will be discussed in terms of enhancing the intellectual depth and quality of academic ecology and of increasing the capacity of academic ecology to answer questions of the sort that are posed in the context of environmental management.

Ecology is the only scientific discipline focussed primarily on organism–organism and organism–environment interactions. It is often assumed that this avowed intention of ecologists, and their efforts over the years, have made them competent to provide immediate answers to all environmental questions. Ecologists are therefore called on during the processes of environmental management: typically, ecological consultants are seen by environmental decision-makers as being conservative in their recommendations, almost to the point of being obstructive. In fact, engineers, physicists and geologists are usually much quicker to supply replies, if not answers, to environmental questions than are ecologists. A brief examination of the subject-matter of ecology will show why this is so.

2 THE SUBJECT-MATTER OF ECOLOGY

There is no way of knowing, *a priori*, which environmental questions may prove to be of interest, nor the diversity of interests which ecology may be called upon to serve. Also, ecologists may need to co-ordinate their work with experts in many other fields. The abundance of a particular species may, for example, be important for medical, aesthetic, geochemical or even patriotic reasons. The management of air, soil and water quality, from the standpoint of public health, agriculture and recreation has important ecological components. The prevention of species extinction, which seems a focally ecological concern, also involves political, economic and moral issues.

Since it is faced with an infinite, or at least non-enumerable, variety of possible questions, ecology cannot provide, in advance, answers to all the queries that might arise. This may sound trivial; certainly the questions of practical interest, requiring mathematical solution, are also non-enumerable. The difference is that mathematics has a finite number of acceptable procedures, which delimit a clearly defined domain. Neither of these statements holds for ecology. Procedures are eclectic, and the domain is too broadly defined to admit convenient formalization.

There is no body of science which can be expected to pick up the leftovers of ecology. A physicist can consider himself finished with an aircraft when the theory is sufficiently well developed to make its flight the domain of the aeronautical engineer, who in turn can hand the finished aircraft to the mechanic and the pilot. Should the aircraft crash or catch fire, it is junk metal. While it may be of professional concern to know why it crashed, the object itself is no longer of intrinsic interest as an aircraft. There is no analogous set of assertions about a landscape. Neither is there a discipline that can take over the responsibility from the ecologists; nor is there a stage of perfection, or degradation, of that landscape that would automatically make it ecologically uninteresting.

Even if there were limitations on the classification of interesting questions, there is a mass of empirical detail to address, which would be overwhelming in complexity if it all had to be considered at one time. There are 2.5 million species on earth. This is a tiny number compared with the number of molecules in a cup of water; however, all the water molecules may be expected to follow the same pattern of behaviour, while each species is more or less unique, often in surprising ways. The problem is somewhat alleviated by the fact that these 2.5 million species interact as 'communities' containing anywhere from 5 to 5000 species. The smaller number is found in hot springs, the larger in temperate woodlands, as well as in tropical forests. Any one of these species, or any subset, may turn out to be of interest for some reason, which may generate a serious question; but in most cases it is not vital to understand all the species or interactions in the community in order to help with a specific problem.

3 ECOLOGY AS AN ALMOST INTRACTABLE SCIENCE

Due to its role in practical decisions and the way its subject-matter has been defined, breakthroughs in ecology may be curiously different from those in other natural sciences and mathematics. Consideration of what might constitute a breakthrough in ecology may therefore be of particular interest from the standpoint of the philosophy and sociology of science.

The goals of ecology are integrative, so that while analytical procedures are important, their value lies in providing the capacity to deal with larger systems. Perhaps some level of intractability is characteristic of all sciences with integrative goals. In contrast, physics is the epitome of a science of the tractable, in the sense that a question which does not fall within the domain of reasonable physical theory can be assigned to that, for example, of meteorology, chemistry or geology. Similar remarks apply, with more or less force, to most of the analytical sciences.

A completely intractable science is in danger of not being science at all. It can maintain its status as science only if it can be demonstrated that it contributes in some way to answering actual questions. Ecology is difficult, but not completely intractable. Since the number of significant ecological questions that may be posed is effectively infinite, stock answers cannot be prepared in advance. The best an ecologist can do is to perform the briefest, and cheapest, *ad hoc* investigation possible, after a particular question has been asked. The quality of ecology can ultimately be measured by its capacity to do so, i.e. the product of ecological research is ecological expertise. The practical measure of an ecologist is not directly his store of knowledge, but rather his quality as an expert in designing procedures for solving *ad hoc* problems that bear a more or less remote resemblance to questions he has previously encountered, or hypothesized.

There are other disciplines whose major significant product is expertise, rather than clear techniques or *a priori* answers. These are all concerned with empirical subject-matter which has not yet been amenable to formal theory. Consider geology, as opposed to chemistry, aesthetics as opposed to geometry, and psychiatry as opposed to neurology.

For most new questions, precisely relevant data have not yet been collected, nor has a precisely relevant theory been developed. This is part of the reason why ecologists are slow to answer questions.

4 NATURAL HISTORY

Ecological systems are historical systems in the sense that particular small events may be magnified by ecological processes, so as seriously to alter an ecological system. Thereafter the traces of the original events may become so inconspicuous as to provide no hint of what had occurred, and no warnings about possible recurrences.

For example, at present in California, Klamath weed, a plant introduced from the Old World, is an innocuous, relatively rare plant inedible to cattle. Fifty years ago Klamath weed was a major nuisance, making thousands of acres unavailable for livestock. This situation was remedied by importing one species of herbivorous beetle that attacked the noxious weed, and has since persisted as a relatively rare insect. Present surveys of the insects and plants of California might show a slight negative relation between the distributions and abundances of the beetle and the Klamath weed, but would show no indication of the importance of the interaction between these two species.

Because we know the history of the region we believe that, if the beetle were to be eliminated, the Klamath weed would regain its dominance, but this could not be determined from a quick survey of the present ecology. Ecologists are therefore concerned about the possibility of subtle interactions between relatively inconspicuous portions of ecological systems that might trigger major changes. This class of possible events also accounts for much of the ecologists' slowness and conservatism. If one chooses more or less to ignore ecology, rapid, courageous replies can be given to environmental questions, but they may prove dangerous.

5 THE CURRENT STATE OF ECOLOGY

Having considered its peculiar difficulties, I now examine the current state of ecology, in the sense of how ecologists may spend their time. They do not use it to analyse the problems indicated above.

Within the discipline there is a subdivision by 'schools' largely based on pedagogical genealogy. The genealogical property is completely lost in such fields as chemistry, and has been divisive in the history of sociology, anthropology and psychiatry. The preservation of academic genealogies is obviously related to lack of formalism and to an excessive breadth of the empirical domain. It may also relate to the goals of the disciplines.

Within each school, workers focus on particular, well-defined sub-areas of research. Each area has a proper domain, while the discipline in which they are embedded does not. Research within ecology may focus, for example, on global biogeochemistry or the micro-environment between individual soil bacteria, and on problems at all scales in between (cf. the table of contents of any standard ecology text). Generally, insights on one scale of observation are of value in interpreting information on very different scales. For example, the fact that clumps of bacteria are anoxic in their centre is of major importance for the global biogeochemistry of nitrogen.

Conceptual schemes range from the mathematical and mechanistic to the sociology and anthropology of man–nature interaction. The pattern of presentation of research results ranges from richly illustrated coffee-table books and personal narratives to abstract mathematics. There are theoretical formulations of such problems as the nature of diversity in ecological communities, flux of energy or elements through large or small natural systems, predator–prey and parasite–host interactions, behavioural optimizations and the interface between ecological and evolutionary events, i.e. there is no shortage of empirical and theoretical research of high quality in the field. The difficulty is that we are aware that the amount that is unknown dwarfs what we now know, and it is not quite clear which paths will take us forward most rapidly.

Ecology preserves its unity as a discipline, despite differences of approach and school, since there is an enormous and undisputed store of information of which ecology is custodian. There are 200 years of observations of organisms, including everything that used to be called natural history, that have no intellectual home other than ecology. (I am extending the definition of the ecologist to include persons that might prefer to call themselves botanists, fisheries biologists, agronomists, etc., in recognition of the fact that they are likely to be called on as ecological experts.)

6 BREAKTHROUGHS CLEARLY REQUIRED

Most published studies conclude with an explanation of what is needed in the future. This permits tentative definition of the more obvious breakthroughs.

As in almost all sciences, many studies end with an appeal for more data. This is not particularly interesting, and may even be self-serving. However, among these appeals some call attention to the need for radically different kinds of data. This often relates to the scale of data collection.

As recently emphasized by Allen and Starr, many of the most interesting ecological phenomena occur on an inhuman scale in either time or space. Slow-motion photography is required to see the feeding activities of many organisms. Fast-motion cinematography is needed to speed up the movements of most plants, and even some animals (i.e. bottom echinoderms) before any attempt at comprehension can occur. Cameras and radios near, in and on the bodies of animals are being used to record locations and movements. These devices are still expensive and are sufficiently difficult to operate that expertise in the handling of the apparatus is more important in their use than depth of understanding of biology.

Microscopes are vital to visualize the life of the vast preponderance of organisms. This has been the case for most of biology for the last two centuries. Procedures for seeing exceedingly large objects are now necessary in ecology. There now appears to exist a general use for 'macroscopes' which would permit us to see enormously large objects. Recently, satellite pictures have revealed phenomena such as the warm-core rings which permit comprehension of the nutrient cycle of the oceans in ways that were simply impossible from shipboard observation. If satellite images continue to be available, similar insights might be expected for terrestrial ecology.

The continuity and distribution of temporal observations may also provide new insights. Standard measurement programmes now usually assume a kind of 'uniformitarianism', i.e. the large-scale or long-term events can be inferred from repetition of short-term and small-scale observation. This assumption may be invalid. For example, the role of the Amazon river in the ecology of the Atlantic may depend more on sudden fluxes of organic material during severe floods and storms than on the kinds of fluxes that may be calculated from routine daily monitoring.

The above examples make it obvious that major breakthroughs may be expected if and when technology for large-scale, small-scale, very short-term and very long-term observation becomes available.

Data handling is now a major bottleneck and can be expected to become worse as new data acquisition techniques are available. Computational speed, and ease of data retrieval and manipulation are of immense importance, considering, for example, that a 5000-species community might have to be modelled as an interaction matrix of fantastic dimensionality.

There is also a desperate need for data manipulation capacity in the context of using data that have been collected *en masse* to serve as base-line information to aid in the assessment of future environmental impacts. These data in published form are curiously dull reading, but are presumably the starting point for interesting analyses, if they could be conveniently manipulated. The problem is accentuated by the fact that many of the relevant data are not published at all, but exist in the 'grey' literature, i.e. reports to agencies, documents in law courts, etc. The relatively simple case of the effect of one nuclear power plant, Indian Point, New York, on one species of fish, the striped bass, has generated several hundred pounds of information-laden paper. Methods of easy access to such data would almost certainly constitute a major breakthrough. At the moment, most of these data have little obvious use, because they have not been collected with the particular question of interest in mind. How relevant are data on fishes in the Potomac when one is concerned with the Hudson?

Even to demonstrate that the vast stores of data collected in the past were truly irrelevant to future ecological problems would also constitute a major breakthrough. At present, computation and access are still sufficiently difficult that they require a person trained primarily in computational and retrieval techniques, rather than in empirical science. Even under optimal collaborative conditions this imposes an intervening communication process between the data and the persons who might be expected to understand them, which makes intellectual progress that much less likely.

Some purely empirical observations may require or permit deep intellectual reassessments of current ecological thinking. In the nineteenth century, when confronted with the suites of strange organisms found in the New World and on oceanic islands, European biologists were almost forced to develop evolutionary theory. The bacterial and animal communities recently discovered at the deep sea hot water vents provide a drastically new and different world to compare with the one we know. Similar remarks may apply to detailed study of anaerobic communities, underground bacterial communities and perhaps planetary explorations.

7 THEORETICAL BREAKTHROUGH MAY OCCUR ON A LESS DRAMATIC LEVEL

The breakthroughs obviously required are those that would facilitate acquisition, manipulation and dissemination of information on a scale that is not now accessible. It is hoped that this, in turn, would permit deeper intellectual breakthroughs.

A persistent effort of current ecological theorists is focussed on finding measures of sufficient generality, so that different-seeming ecological systems can be shown to be following similar rules. Species diversity studies, predator–prey models, and simplified indices for community description models all have stimulated this set of endeavours. What may help are measurements which are information-rich, so that we can make use of all past efforts, and at the same time theoretically neutral, so that our conclusions are not prejudiced by our measurements. A breakthrough in measurement would consist of finding a measure which is not only information-rich and theoretically neutral but also useful for the formulation of theory in the way that colour and brightness were useful in the development of main sequence theory in astronomy.

8 HOW CAN INTELLECTUAL ADVANCE IN ECOLOGY BE ENCOURAGED?

At the scale of this article, it has only been possible to describe the format of ecology, without actually presenting detailed examples. Obviously, intellectual advance is not merely a change in format. A generic sense of how to recognize incipient intellectual advances is the artistry of a first-class scientific administrator; it is not a science in itself. I have some suggestions. They are personal, tentative and applicable to most sciences, to a greater or lesser degree. I do not believe that the entries on this list are logically independent nor could I rigorously defend them. They do relate to my own research style.

First, I believe that questions about organisms or specific features of particular environments are more likely to provide a breakthrough than questions about higher-level theoretical constructs. This is almost the reverse of the situation in physics, in which the manifestation of basic entities is easily accessible, while the entities themselves are obscure. In ecology there is no doubt of the existence of individual organisms and of the biosphere itself. Populations, species, communities are theoretical constructs. Stability of populations, for example, is a higher-order theoretical construct whose empirical meaning is not perfectly clear. Empirical statements about organisms or about the entire biosphere may therefore prove more exciting than inquiry into new statistical procedures or modifications of theoretical models.

Secondly, questions which do not admit of easy verbal formulation may be the most interesting ones. Once we can name things clearly, much of the intellectual work is complete. Often, this type of question arises from novel kinds of experiments; for example, the controlled perturbation of the Amazonian rain forest is being done on a uniquely large scale. Can the results of this study shed light on micro-habitat studies? Specifically, can a perturbation on the scale of tons of trees and thousands of acres be shown to follow the same pattern as perturbations on the scale of milligrammes and centimetres in, say, a community of soil arthropods and moulds? What, if any, conceptual patterns are interesting in both contexts? At present, even the terminology for this kind of scale transition is non-existent.

Thirdly, formulations which fall between present theoretical subfields are more promising than those that are clearly assignable to subfields. For example, there is a polemical literature related to biogeography that focusses on statistical evaluation of precise predictions in terms of imprecise data derived from field collections. It is certain that field distribution data will remain somewhat imprecise. Perhaps the only resolution of the controversy will lie in a theoretical formulation that accepts the existing predictions about field collections, but also makes predictive assertions about other data sets which are relatively free of statistical ambiguity. For example, if a theory predicts biogeographic distribution and also the morphology or physiology of particular organisms then the polemics might be ended.

In general, intellectual breakthroughs can occur only by an analysis of the truly unknown, rather than by obsession with the imperfectly known.

9 CONCLUSIONS

Unlike engineering, in ecology there is neither a unique object of study, a unique purpose of study, nor a unique technique. Nevertheless, we do have the advantage of very real empirical questions, although we cannot answer them very well without *ad hoc* investigation.

While some ecologists are too well aware of their own weakness in providing answers, others, perhaps lacking sufficient knowledge for humility, will provide whatever may be needed as certification for decision-makers. This can be dangerous.

Ideally, ecology should be able to provide relatively cheap, quick and valid *ad hoc* investigations of practical questions by virtue of its backlog of natural history information, and current theoretical and technical capacities. In fact, while we are better at what we do than non-ecologists, our information backlog is neither yet rich enough nor sufficiently well organized and accessible to be as useful as we should like.

There exists a real possibility that more data, with more convenient access, may not really help us. It may be that the collection of data must be modified to make it more specific and to demonstrate this would also constitute a major breakthrough. Our data-collecting techniques are not adequate to our needs, and our data-processing procedures form a bottleneck. These problems are primarily technological but they are fundamental to more intellectual breakthroughs, whose present form can be only dimly discerned.

The generalizations of ecology are best tested in terms of the capacity of the discipline to deal with questions and problems generated outside the discipline. Our situation is, in a sense, comparable to that of eighteenth-century medicine, in which the wise patient ought to have chosen his physician by his record of cures, rather than his theories. In a sense, the quality of ecology is to be measured by 'clinical' standards.

Intellectual, as opposed to technological, breakthroughs in ecology will come from fundamentally new approaches, rather than from refinement of our present simplified models. This is, of course, valid for most sciences.

Public knowledge, truth and the ways of scholarship advancement in western civilization

Concluding remarks

S. N. EISENSTADT

I THE WESTERN VIEW OF KNOWLEDGE

REFLECTING on the views and comments that have been presented at this colloquium has led me to consider how they would have been received by traditional thinkers from the Far East (as typified in some of our western literature). I concluded that they would have been rather bewildered, not by the fact that we were engaged on different things (not very practical things in the narrow sense) nor that as some of the patronage funds we had been receiving rather abundantly of late were being cut down, we felt that something should be done. They would understand this. They could also easily understand some of what *we* call the scholarly disciplines – mathematics, for example. I think, however, that they would have been bewildered by some of the presuppositions running throughout our discussions.

One presupposition – a very strong one – was, in a sense, inherent in the topic of the colloquium, namely that all these different types of not-so-very-practical activities (even though they may have practical applications) are bound together by a common thing which we call 'knowledge'. Perhaps they would still have accepted this point, but they would understand much less our very strong emphasis on 'truth'. They know it, of course, but to talk about 'truth' in general, and especially to speak about art or the natural sciences and use the word, seemingly with the same meaning, would be confusing. They would be even more confused when we talked about 'progress', 'advance' and 'generation' of knowledge. They would understand when we said that we know more of some things than others but the general concept would be strange. They would understand that people engaged in our type of activity need patronage and that quite a lot of this patronage comes from rulers, and they would understand perfectly, in China at least, that most of these rulers would be bureaucrats. But I do not think they would really understand what it means – to use the title of one of Professor Ziman's books – by this knowledge being *public knowledge*, something that the public or its representatives are somehow involved in and can judge. They would not understand that part of the accountability to the patrons is directed towards a general public.

Yet we have obviously made all these suppositions, so what we are doing here is related to some basic suppositions about our culture. Indeed, unless we take these suppositions into account, I do not think that we can really fully understand what we are doing at all.

Obviously, we are worried about the possibility of the continuation of the pursuit of scholarly work due to a shortage of funds. There is also a certain unpleasant public atmosphere, distrust of scholarly, scientific activities compared with the trust that existed before. But, beyond these, there is something much more profound.

Let us, first, look at two very crucial institutional aspects which we take more or less for granted. We are financed, to a very large extent, by some centralized government or by semi-governmental agencies. Foundations, public bodies and many institutions still have their own endowments but, even these, are more and more related to government. This strong dependency of learning, or the pursuit of knowledge, on governmental funds is a new development, even in western society, and this naturally results in very strong competition between the different institutions applying for grants.

Of especial importance is the fact that this pursuit of knowledge is carried out in institutions which, to some degree, are part of our educational institutions and the educational system. This means that we assume that the transmission of knowledge is very strongly connected with innovation, with breaking new ground and with the search for innovation. This would have seemed incomprehensible to my eastern friends. Education, yes; practical knowledge, yes; speculation, yes. But if you combine these under the canopy of widespread educational institutions and *stress* innovation and change in knowledge, this would have been regarded as curious, perhaps even inhuman. Similarly, the idea of public knowledge would have been alien. Today we have a growing democratization of this public knowledge, although, as I think Professor Ziman points out, this could already be discerned very early in modern science. The important change is in a true educational change. One cannot compare the eighteenth or early twentieth-century public to that of the present day. And I do not think we can understand the problems we are discussing if we do not take this new, wider public into account. We thus have a new institutional reality which is very forceful, and which has grown out of certain cultural assumptions.

2 CULTURAL ASSUMPTIONS

What, then, are these assumptions? How can we explain this specific attitude to knowledge, innovation and education that is characteristic of western but not of Chinese, Indian, Buddhist, or even Islamic civilization?

The clue lies in the very unusual relations existing between the concepts of knowledge and those of salvation – if you will allow me to take up a term which has been used by Max Weber, one of the founding fathers of my discipline – which have characterized western civilizations. Through the whole history of western civilization – starting in the first encounter between Greek culture, Judaism and Christianity – there has developed a stress on the pursuit of knowledge combined with a stress on salvation. Weber pointed out the central importance of the idea of salvation and how the different approaches towards it provided an important clue to understanding the different civilizations. Salvation is, of course, a Christian word, but he used it across the board, with respect also to other civilizations. We find already in the medieval universities, a combination, uneasy but very fruitful, of the quest for salvation and the pursuit of knowledge which became fully articulated in the seventeenth/eighteenth centuries. In the seventeenth, eighteenth and nineteenth centuries, the pursuit of scholarship, of science and of truth, became something which combined different human intellectual activities and constituted a very fundamental platform of western culture. It grew out of the assumption not only that such activities were useful in the narrow sense but that they provided the mix from which a civilization and its special mode of life was constructed. A consequence was the creation of a very curious tradition, one which stressed expansion, innovation and change.

At the same time the idea of 'public' knowledge started to develop. This concept stressed that knowledge was not to be confined simply to groups of esoteric specialists in small, sacred gatherings or to craftsmen. Knowledge became increasingly public, more and more part of the general culture, of the general 'formation', and slowly became institutionalized in universities and other institutions. New disciplines were accordingly created: physics which we talked of as a sort of 'model' discipline, biology, history and the beginnings of the social sciences, sociology, political economy and so on. They were all gradually brought into the institutional area, very often in different ways and, as mentioned by Professor Nowotny, after many struggles and degrees of institutionalization.

What were the presuppositions which resulted in these different disciplines coming together? What was the common factor? It was not only the pursuit of truth (knowledge) in a general way, but also a combination of the theoretical approach with the analysis of empirical facts. Mathematics was a partial exception but, as Professor Atiyah has pointed out, the distinction is less and less evident now. When we talk of truth, we usually mean some kind of a combination involving connections. This is *our* meaning of scientific, scholarly truth, which may not be exactly the same in other civilizations. In our case, the general underlying idea is that we combine theory, conceptualization and reconceptualization with some sort of empirical testing. I have deliberately used the phrase 'some sort' because there are different ways of conducting such testing, but in all there is some combination of empirical testing with the different modes of looking and of theorizing, although philosophy and more especially theology do not really fit this pattern. All the other disciplines posed this problem of how they could combine the transmission of knowledge with innovation and the broadening of new frontiers of knowledge. All seemed to be moving in this direction and in the nineteenth and early twentieth centuries everything was seemingly beautiful. I am, of course, exaggerating because already there were reactions against the trend, but on the whole, I do not think that I have distorted the picture.

3 RECENT CHANGES

Great changes, however, have taken place in the last 20 or 30 years, and we must ask in what way exactly. I am not referring simply to the administrative situation, although this too cannot be fully understood unless some account is taken of the broader institutional–symbolic problem.

First of all, there was, of course, the great growth of knowledge. The process which has been illustrated for physics has also taken place (in lesser and different degrees) in other disciplines. When one compares what is known today, even in sociology and economics, with the situation that existed 70 years ago, the sheer growth of knowledge is stupendous. As mentioned by Professor Ziman, this has also led to growing specialization and a growing division of labour within all the scholarly disciplines. Without funds, this growth could not have happened. The sources of money have, moreover, became more and more public, less and less private.

Then, too, through technology the impact of science on ordinary life has been growing. Knowledge, or at least parts of it, is changing not only our conceptions but also our real lives. Whether the percentage of science related to such changes is 10, 20 or 80 per cent is a moot point, but the accepted image is that the whole of science has been influencing our lives, giving great impetus to new possibilities. Together with this has occurred, of course, the spread of education.

More recently, problems have appeared which are concerned with the very nature of our existence. The atomic bomb posed the question of ethical responsibility in a much sharper way than previously, as we have heard. This is now a subject of intense debate in medicine. But until some 15 years ago there was general optimism that we should be able to cope. This belief was closely related, particularly in scientific circles and partially among the increasingly educated general public, to the development of an image of science which was never fully explicit but was somehow very pervasive. We have heard a great deal about this in our recent discussions. What was this view? First of all was the notion of the great power of science, this cumulative combination of the deductive, inductive and experimental approaches. A further element was that science was becoming increasingly predictive, closely related to which was its mathematization, its quantification: the more quantified, the more scientific, the more predictive, and so on. Finally, the view developed that science had many potentialities of application, that it could even give prescriptions for the solution of many social and political problems. Scientists themselves succumbed very easily to the temptation to believe that they could give prescriptions for so many problems.

Then came Thomas Kuhn, who had a very interesting and rather contradictory impact on the public perception of science. On the one hand, he seemingly demystified it because many people understood him to say that science was no better than any other esoteric knowledge. People believe in something, they do not really test it, and then something new comes to displace it. But in the popular mind, there was also a different impact, namely that paradigms *do* indeed change, but that they change for the better. A more powerful paradigm comes along and supersedes its predecessor. This view has had a very powerful effect on the image of science in the public mind – an image which was also reflected in our discussions – and which, as I shall try to show, is not entirely accurate. But it did affect the image of scientists, the importance of scientists, their own self-perception and the public reaction to science. This view became connected with the great increase in financing, the growth of – to borrow a neat expression of Professor Ziman – 'the collectivization' of science, or its bureaucratization. Gradually, or perhaps not so gradually, all these developments have given rise to a growing disenchantment with science.

4 A DISTRUST OF SCIENCE?

This situation did not arise solely because of the economic recession, without which the problem might have been slightly, but not entirely, different. Some problems would still exist, notably that science could be invoked to provide any number of prescriptions for the ills of society or of mankind. In the social sciences, many of the prescriptions eagerly proposed in the 1960s for problems of urban renewal or desegregation turned out to be rather short-sighted; in economics, they failed to anticipate the problems of stagflation, and there are similar examples in the attempts at the application of the natural sciences to broader human or social problems.

The growing awareness of this situation – while the scientists were often still pretending to have definitive prescriptions – naturally gave rise to a strong backlash among the broader public and its various representatives including politicians, public administrators and administrators of foundations. A degree of disenchantment with science and scholarship was generated, a distrust about which we heard much in the 1960s and 1970s. This backlash was also connected with intellectual trends undermining the image of the scientist. All this occurred alongside the bureaucratization and the growing dependence on governmental or public agencies of various types. In this context, a far-reaching change has also taken place in the relations between the educated public, the administrators, the patrons-to-be and the scholarly communities. It seems to me that we are not yet fully aware of the extent of that change. We *are* aware of the distrust (which may now be simmering down), but I am not sure that we are sufficiently aware of the fact that, in the meantime, a new type of structure has developed in the relations between scientists and funding agencies, a structure which has a far-reaching impact on the pursuit of scholarly work. A conflicting situation, almost a contest, has arisen between the patrons and the receivers of funds for research, and it constitutes a far-reaching institutional change. The interaction has become less and less one of mutual exploration and support and more and more one of confrontation and contestation. This is built into the structure, into the way in which the funding is organized, so that, wittingly or unwittingly, the agencies have an extremely powerful impact on scholarly work.

Whereas the patrons of science and of scholarship and the scholarly community may share the concept that the main objective is the search for knowledge and for truth, the administrators, the politicians, still hold on to the image of science as a prescriber. The scientists cannot live up to this image but are afraid to admit as much. They find it difficult (as mentioned by Professor Nowotny) to abandon the role of the expert and still try to emulate the image I have described.

This new situation is much more profound than is evident from administrative data. We have an institutional structure into which distrust is built and this has an impact on the whole scholarly community, even if some disciplines may be less touched by it. Mathematicians are probably the least affected; the historians rather more. In Holland it has been decided to abolish two chairs of Sanskrit and similar studies, which shows that even historians and philologists are caught up to some degree and, in percentage terms, perhaps to the same degree as others. How should we react? What have we learned? What have I learned from this seminar? One answer is that I have learned much more than I expected, and I am extremely grateful.

I did not learn that the distrust of science is really justified, but I did learn that the distrust of that image of science, which I have presented and which is a very popular image, must be reckoned with. We have to present ourselves in new ways, but, above all, to *think* about ourselves in new ways. First of all, we still have that very strong common core which is at the centre of western civilization, namely the pursuit and advancement of knowledge, of truth, based on some relation or interplay between conceptual, theoretical and empirical references, some testing of deductions. This, I think, is still common to almost all our disciplines.

Because of this, we still believe in that scholarly community, in universities and research institutes, where public knowledge makes progress on account of some combination of accumulation and reconceptualization. Indeed, on the basis of our discussions, the definition of scholarly breakthrough is not so different across the different disciplines even if the details vary. There is a greater or lesser fragmentation, more or less division of labour and different timespans of activity. Marc Bloch's book, which has been mentioned, I know very well and I fully agree was a great breakthrough. Why? Not just because it taught something new, not just because it was a very powerful reconceptualization, but because it combined such reconceptualization with an analysis of very rich data. And this, I think, is the same everywhere.

The specific details of the way in which a small piece of research can be translated into something of broader significance varies with the discipline, but in essence it is similar in all. There are differences, of course. For example, the greater the division of labour, the less perhaps can any single work be understood directly. But, basically, when we talk about the same type or mode of knowledge and the image of breakthroughs, we are sharing the same presuppositions. And, because we share them, the nature of advance is relatively similar across the disciplines, even though not in content, in tempo or in organization.

While, however, we share these presuppositions about knowledge and breakthroughs in knowledge, it should be very clear that the methods of achieving them are not the same. Not all sciences have to be quantified in the same way or to the same degree, and the time-lag between accumulation and reconceptualization differs from discipline to discipline. Variations are to be found also in the degree and nature of their applicability.

There is no single formula that applies, and even in any one discipline the paths are multiple, as Professor Dik's paper on linguistics and Professor Malinvaud's on economics have clearly shown. There is no single omniscient paradigm and in consequence there cannot be a single prescription. And we should be absolutely clear about this. There may be good, or better, or worse solutions to different problems, but there is no single prescription.

What brings us together is the search, the type and the methods of the search, but not the concrete relation between accumulation and reconceptualization and the time-lags associated with applicability. We should more and more present the differences honestly and pretend much less that there is one knowledge (or something of the kind) which is exactly the same everywhere. This notion may appear to be quite trivial, but I think many scholars and the public have forgotten it. And because in the past scholars accepted the public challenge to present themselves as sages, the public has to some degree become distrustful of them and, I think, we have to change this.

At the same time, we have to stress that our major function or mission is not so much to give prescriptions, but, I strongly believe, to broaden the range of public discussion. This applies not only to the humanists and the social scientists, but to all scientists. Providing alternative ways of perceiving problems can be much more important than – to use a caricature – 'selling prescriptions'. I think that, when we took upon ourselves the role of gurus, we forgot this very important task of broadening the range of public discussion in a constructive – perhaps controversial, but constructive – way.

5 SUPPORT FOR SCHOLARSHIP ADVANCEMENT

This brings me to the last point with which we have been concerned here, the most mundane aspect of the institutional complex of growth of knowledge. How do we apply for funds and identify advances? Identifying advances *a priori* is very difficult, almost impossible. What we can do is to see what the conditions are which may stultify advance, and work around these conditions and change them. I wish to refer again to the type of report which many grant-giving bodies demand. They give you money for a year or two, but you have to state exactly, in the research proposal, what will be the outcome. We all know this. The procedure, although it probably produces quite a lot of good work, tends to shorten the time-span, the breathing space for many scholars. And I think this is bad. It may be unavoidable to some degree, but I think that more and more counter-institutional ideas are needed. Some of these have been mentioned by

Professor Ziman, and there are many others. I should like to quote one example from the Research Council of Canada which offered a grant (I do not know whether it still exists) to groups of scholars, whether in the same institution or across the country, for a relatively innovative project over five years. They judged it partly in the same way as in Sweden, by interrogating the people and by considering their reputation and the novelty of the proposal.

There are risks, of course, but some risks must be taken; we simply have to spread them out more and to loosen, to a very high degree, the time constrictions which, I think, are very detrimental. We also have to be more and more open to the possibilities of interdisciplinary approach. Here, I should like to be very cautious, because, in my experience, when one talks of 'interdisciplinary', there is often a lot of 'inter' and very little 'discipline'. One should beware this. There is no doubt, however, that, at the same time, it can be extremely fruitful. I have, myself, learned much from historians, for this is my type of sociology; others would learn from other people. We need to open up more, to encourage more interdisciplinary work, with wider time-spans.

I should like to go beyond these relatively simple institutional aspects and return to the point I have already made, namely that we have to change our relations with the public and try to incorporate the public increasingly in our work. We must find different ways of doing this, which will, of course, vary in different disciplines and in different institutes. We must look upon members of the public not as adversaries but as *participants* in the common endeavour. In my university's appointment committee, the members from the public do not sit there as representatives of a board which gives away money. They see themselves as members of the university. They are interested in what we do and are perhaps as committed as we are ourselves, even if, naturally, they have different views on many problems. At the same time, they are an important link with the broader public.

I think we have not been sufficiently aware of the weakening – I will not say disappearance – of the influence of that enlightened public, which feels itself to be a part of the scholarly enterprise or, at least, willing to support it, that has come from the growing dependence on governments and on central agencies. We have to show institutional imagination to bring them back, to create new structures and frameworks. And this will allow us, I think, to ease some of the administrative problems which I have mentioned previously, and which in fact reflect wider, more important problems because they really relate to some of our basic concepts of the way we look at knowledge as a part of our civilization.